£3.50

CLEARED FOR TAKE-OFF

CLEARED FOR TAKE-OFF

MEREDITH HOOPER

Illustrations by
TONY TALIFERO

ANGUS & ROBERTSON PUBLISHERS

*Unit 4, Eden Park, 31 Waterloo Road,
North Ryde, NSW, Australia 2113, and
16 Golden Square, London W1R 4BN,
United Kingdom*

*First published in Australia
by Angus & Robertson Publishers in 1986
First published in the United Kingdom
by Angus & Robertson (UK) Ltd in 1986*

ISBN 0 207 15233 0 paperback.

ISBN 0 207 15474 0

*Typeset in 12 pt Baskerville by Setrite Typesetters
Printed in Singapore*

CONTENTS

ACKNOWLEDGEMENTS

Individual members of Qantas staff in Sydney and London, and members of Qantas crews on several England–Australia flights, made much of the research for this book possible. I am most grateful. The Department of Transport in Australia gave me access to air traffic control; and the Librarian of the Royal Aeronautical Society in London access to the Library's resources.

I would like to thank Captain Alan Terrell, Captain David Long, Peter Elliott, Michael Tarrant and Michael Cottee of Qantas and James Mayfield of the Australian Department of Transport for particular help.

MEREDITH HOOPER
London, 1986

PICTURE CREDITS

The publisher would like to thank the following organisations for their valuable assistance in the preparation of this book.

Information and pictorial reference material for the illustrations were supplied by: British Airways, Sydney; Marriott In-Flite Services Ltd, Feltham, Middlesex; Qantas Airways, Sydney; The Science Museum, London.

The photographs were reproduced by kind permission of: Austral-International Press Agency and Photographic Library, Sydney (cover background); Australian Picture Library, Sydney (p. 80); P. & P. F. James, Hounslow, Middlesex (p. 88); News Ltd, Sydney (p. 44); Qantas Airways, Sydney (front and back cover, half-title page, title page, contents page, pp. 1, 5, 9, 14, 19, 25, 51, 80); The Singapore Tourist Promotion Board, Sydney (p. 59).

Special thanks go to John Squire of Qantas, who took a number of the photographs that appear in this book.

AUTHOR'S NOTE

Measurements in aviation are a mixture of imperial, nautical and metric. Runway lengths are given in metres. Meteorological information states visibility in kilometres, cloud base in feet. Altitude is always given in feet. Technical crew think, and talk to each other, in nautical miles. Distances on the flight deck are expressed in nautical miles, and speed in knots (nautical miles per hour). Nautical miles relate directly to the spherical surface of the earth and the way it has been divided up into latitude and longitude. One hundred nautical miles equal approximately 115 statute miles and 185 kilometres. The maximum speed of a 747 is around 542 knots, 625 miles per hour, 1000 kilometres per hour.

People in aviation need to work to exact measurements, commonly understood. Following aviation practice, altitude in this book is given in feet, distances as flown by the 747 in nautical miles (written as 'miles') and the speed of the 747 in knots.

PROLOGUE: LONG-HAUL FLYING

Take a kangaroo. Swell it to five times the size of any living kangaroo and make it leap the height of a five-storey building. At the top of the leap, fix it permanently.

Take a hot day in summer. A Boeing 747, engines turning over, waits at the head of runway R34 at Sydney's Kingsford Smith Airport, for control tower clearance to take off. The kangaroo, painted on the tail of the 747, is over 45°C to touch: aircraft metal heats up on hot days at airports. Clearance to take off, and the 747 turns into the runway, moving at the speed of the family car on a suburban road. Three hundred and thirty tonnes of metal, fuel, humans, cargo, food and water. A complex machine of four and a half million separate parts, as long as six buses, wider than a six-lane motorway. In 30 seconds it is travelling at twice the motorway speed limit. Ten seconds more, the nose tilts up, and the 747, travelling at 160 knots, lifts off from the runway and climbs at an angle of 10 degrees. The temperature of the kangaroo is already dropping as, at 250 knots now, climbing steadily, the air streams past the giant aircraft. Nine minutes after take-off the 747 is 10,500 feet up in the sky, and travelling at 350 knots. The air temperature has dropped 30°C. The kangaroo is still cooling.

Ten minutes from take-off the 747 receives instructions to turn on course to Singapore. The kangaroo has travelled 45 miles. Twenty-five minutes later the 747 achieves a cruising height of 35,000 feet, and a speed of 500 knots. The kangaroo is moving above planet earth at 85 per cent of the speed of sound, and is 6000 feet higher than the highest point on the earth's crust — the icy peak of Mount Everest. The temperature of the air is minus 50°C. Colder than almost anything on earth.

CLEARED FOR TAKE-OFF

The Boeing 747 — the "Jumbo Jet" — was the world's first wide-bodied airliner.

At 35,000 feet the 747 flies above most cloud. But there can be incredibly strong upper atmosphere winds travelling at 200 knots. On earth, winds reaching that speed would be destructive typhoons. Up here, in the tropopause, on the edge of the stratosphere, aircraft can use these jet streams — catch them like a tide — to cover distances even faster. Or the winds can buffet the plane, reducing speed. Up here the painted kangaroo can be hit by hailstones, or be shaken by 'clear air turbulence' which does not show up on radar.

If the painted kangaroo could see, it would see continents, islands, oceans below, and the universe all around and above. If it needed to breathe, it would gasp, because here there is only one-fifth of ground-level oxygen. If the kangaroo's outline had any living thickness it would swell because the pressure of the atmosphere at 35,000 feet is between one-third and one-quarter of the pressure of the atmosphere on earth.

Inside the 747, people sleep, eat, work, read, watch films, go to the lavatory, walk around, drink, chat, listen to music. There is some noise, some vibration. The air is changed every two minutes so smells don't linger. The temperature is set at 22°C. Air pressure inside the cabin is kept the same as it would be at 8000 feet — like being nearly one-third up Mount Everest, or a little above Australia's highest mountain. The speed reached at take-off — 160 knots — is faster than most people will ever have travelled along the ground. Yet, strapped into a seat high up above the runway, higher than the roof of a single-storey house, 160 knots does not feel so very fast. And, at 500 knots, it is possible to forget the sense of speed completely.

The painted kangaroo will travel halfway around the earth in 24 hours. It will come down out of the skies just twice before reaching London, on to two small islands, Singapore almost on the

equator, and Bahrain in the Arabian Gulf. At Singapore the evening is humid and warm. At Bahrain in the still middle of the night it is cool and dry, before the desert heats up at dawn. The painted kangaroo will travel 17,708 kilometres, 11,003 statute miles, much of the time in darkness.

As the 747 heads north and west out of Sydney, body time and clock time stop synchronising. Singapore is reached when, for Singaporeans, it is 9.15 pm in the evening. Tired cleaning ladies wait by their brooms for any litter, then whisk it away. At Bahrain, seven and a half travelling hours later, it is 1.00 am in the morning. Sleeping Arabs in striped gowns lie stretched out on the air terminal benches, heads resting on briefcases. Up and away from Bahrain the sun begins to rise over Russia. In the grey light the earth looks frozen. Snow covers the Alps and lies in drifts across Germany. The air temperature is minus 41°C. But as the 747 begins a long, slow descent the temperature gradually rises. At Heathrow Airport it is 0°C. The aircraft is 22 hours three minutes' flying time from Sydney. Body time says it is late in the afternoon, but it is 5.55 am, and barely light. Airport personnel stamp cold feet in the freezing air. It is the middle of winter.

The painted kangaroo has crossed the world. It has come from summer to winter in less than a day. It has telescoped time so that the day is just beginning when it should be near its end. It has hurtled from standstill to near the speed of sound, from the earth to 35,000 feet above. Its surface has experienced changes in temperature of 100°C. Its structure has withstood a built-in headwind of 300 knots as the airflow has streamed past its nose. Yet it arrives as serenely as it left. A painted kangaroo, in permanent leap, on the tailplane of a Boeing 747.

And in 14 hours 20 minutes' time it will be doing the same thing, all over again, back the other way.

CLEARED FOR TAKE-OFF

Every day two million people fly as passengers on the world's airlines.

1. PASSENGERS

Wilbur and Orville Wright built an aeroplane out of wood and wire, linen and glue. The pilot lay face down on the lower wing, left of centre, to balance the weight of the little homemade engine. With its small tin petrol tank, hand-carved propeller and sledge-like landing skids, the Flyer *weighed about the same as* four men. It flew for the first time on a freezing cold, wintry morning, on Thursday, 17 December 1903. It flew a distance little more than half the length of a Boeing 747 and as high as its underbelly, darting and lunging about into a heavy wind, and remained in the air for 12 seconds.

Wilbur and Orville Wright experimented with flying at Kitty Hawk, North Carolina, on the American east coast because there were good strong winds blowing in off the Atlantic ocean, that helped to give the aircraft lift, and because there was sand to crash into, and very few people around to interrupt them, or fall on. The *Flyer* took off around half past 10 on the morning of the 17th because the brothers' testing and experimenting and rebuilding had reached a point where a flight could work. A strong wind was blowing, five witnesses had come to watch, and the brothers had promised to leave Kitty Hawk and be back home in Dayton, Ohio, for Christmas.

A Qantas Boeing 747 leaves Melbourne for Sydney, Singapore, Bahrain and London every day at the same time because the schedules say it will, and hundreds of people expect to travel on it; because the world's airline companies have negotiated amongst themselves and a time slot has been allocated for a Qantas 747 to land at Heathrow Airport before 7.00 am every morning; because there is aviation fuel for the tanks, and the correct number of meals in the galleys; because there are airport staff and air traffic control personnel available to process the flight off the ground and through the skies, and technical crews to fly it, and cabin crew to look after the passengers, and sufficient drinking water in the three tanks at the back of cargo hold 2.

No-one took much notice when the Wright brothers managed to get their heavier-than-air machine to fly in 1903. For the next two years Wilbur and Orville practised taking an improved aeroplane up into the sky in a field near their home in Dayton. Through hundreds of attempts, through crashes and amazing near-misses, bruised ribs and bloody noses, through carefully building up their understanding and experience, Wilbur and Orville taught themselves how to fly. Measuring distances flown, time taken, speeds reached — and breaking records which they themselves had set because they were still the only men in the world who could fly — they believed in the future of the aeroplane. It could be used for exploring, they thought, for discovering what the enemy was doing in war, for sport or maybe even for carrying mail.

*On 17 December 1903 the Wright brothers' **Flyer** made history with the first successful powered flight.*

Wilbur and Orville returned to the lonely Kitty Hawk beaches in 1908 to test an improved aircraft. Instead of lying face down on the wing the pilot could now sit upright on a small seat. There was a second seat for a trainee pilot or a military observer. On the morning of 14 May Wilbur told the young mechanic who was helping them, Charlie Furnass, to climb aboard, and took him up for a flight lasting half a minute. Then Orville took him for a longer flight lasting three minutes 40 seconds. Furnass felt the machine lift and saw things down on the ground blur past, then become clear as he rose higher. The engine just behind his back roared deafeningly, the wind hit his face so that he gulped for breath. Beside him Mr Orville moved a lever, the right wing rose and the machine swung round in a turn, yet he was still sitting in his seat. He was not flung out. He felt safe.

Charlie Furnass was the world's first aeroplane passenger. The aeroplane had proved that it could lift two humans into the air at once.

One Boeing 747 can carry 487 passengers and three tonnes of mail and cargo one-quarter of the way around the world without

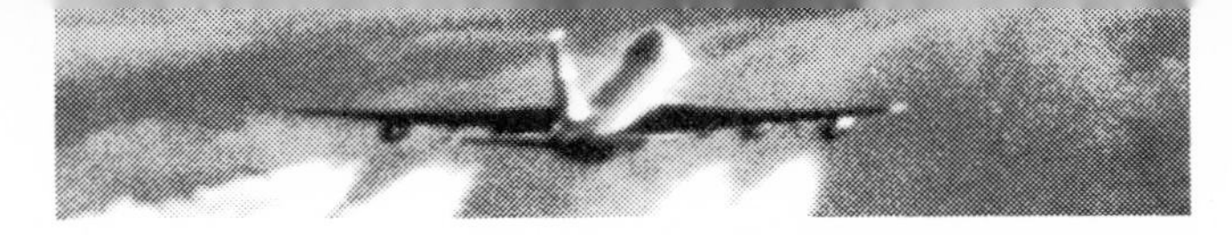

landing, if it has to. A small boy, who happened to watch the strange wood and fabric biplane of Mr Wilbur and Mr Orville swoop over the trees in Dayton, could have flown in a Boeing 747 as an elderly man. The airline business belongs to our century. The history of passenger flying in powered aeroplanes can still be contained within the span of a single lifetime. Every day two million people fly as passengers on the world's airlines. Flying has come from a single aeroplane passenger to three-quarters of a billion per year in three-quarters of a century.

Almost all the business of running an airline happens behind the scenes. Passengers are a major reason for the intense organisation and activity, but they do not see it. Even the first flying machine had to have a place from which to take off and somewhere to land, someone to control it and someone to make sure it was working. As the tens of people who were willing to travel in aeroplanes turned into hundreds, then hundreds of millions, aircraft became bigger, faster, more complicated and more efficient, and the organisation to make sure they departed and arrived, were controlled and maintained, has grown vast. A passenger aircraft leaving on a scheduled flight today is part of a tightly woven network stretching around the world, and forward and back in time. Once having taken off, the aircraft is part of the organisation of the air, that envelope which surrounds our earth, filled with winds, temperatures and pressures, and other flying objects.

2. AIRCRAFT

What happens on one flight on one night in the airline business may or may not happen the next. What happens in one month might be changed the next. Commercial airline companies work to carefully stated procedures and precise timings, with constant repetition. Yet each flight of every aircraft is a unique event — a particular combination of aircraft, crew, fuel load, weather and temperature, time and place, nature and weight of the payload. But if the whole complex process is sliced through, dissected along the length of one flight on a scheduled service, much of what is found is common to most airlines most of the time.

This is the biography of one aeroplane, VH-EBQ, a Boeing 747B, and its long-haul flight between Australia and England, departing on a Friday in May. It is the story of what has to happen to make the flight possible — the world behind the apparently empty aircraft standing on the ramp, with seats ready to be filled by passengers. It is the anatomy of a route on which it now takes less than 24 hours to cross the world, telescoping day into night, and reversing the seasons.

Qantas Flight One is scheduled to leave Australia every day at the same time. Airlines give their routes shorthand names, used by booking agents, on tickets, in airport terminals, and by controllers and planners, local and international. Qantas has assigned every scheduled service, out of Australia and return, a pair of QF numbers. QF13/14 is not used, in deference to superstition. 'QF1' stands for the daily flight from Melbourne through Sydney to London. It is the concept of the flight, the idea of it, the promise.

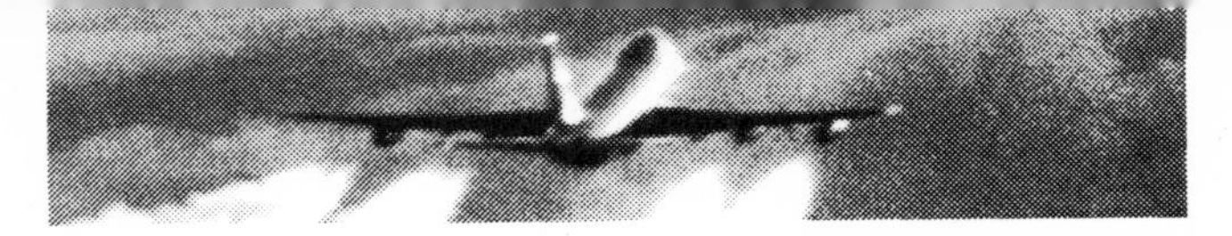

QF2 is the same journey, return, leaving London every evening. British Airways call their flight along this route 'BA12' leaving Australia, and 'BA11' from London. The Qantas flight appears as 'QF001' when generated by a computer. And over France it is 'Quebec Foxtrot One', because French air traffic controllers say they cannot pronounce 'Qantas' so use the international aviation code, with an agreed word standing for each letter of the alphabet.

The aeroplane which is flying this particular Qantas Flight One, leaving Melbourne at lunch time on Friday, Sydney mid-afternoon, and arriving in London early on Saturday morning, is Boeing's 410th 747 and the 19th bought by Qantas, VH-EBQ *City of Bunbury*. In the last five days it has flown to the Philippines and back, across the Pacific via Honolulu to San Francisco and return, up to Manila and on to Hong Kong, returning only two hours and 25 minutes before departing this Friday lunchtime as QF1. Waiting in Melbourne before taking off, it has flown exactly 10,125 hours 30 minutes, and has made 2090 landings.

Aeroplanes do not earn an airline any revenue sitting on the ground. From Sunday to Friday VH-EBQ has served as QF19, QF20, QF12, QF3, QF4, QF27 and QF28. But for the week before that it sat idle in a hangar at Sydney's Jet Base having four new Rolls-Royce RB211-524D4 engines fitted.

Airlines buy aircraft and engines separately and the competition between engine manufacturers to sell their powerplants is intense. All Qantas 747s were equipped with American Pratt & Whitney engines until the decision in 1978 to buy Rolls-Royce. The D4 was being developed, promising to perform with less fuel, and more quietly, offering a saving of around $1.4 million a year per aircraft in fuel costs, a higher take-off weight and longer range flights. New engines have a lengthy gestation before they are ready to come into service. In the interim, VH-EBQ was fitted with the rather less efficient Rolls-Royce RB211-524B2 engines when Qantas took delivery of it from Boeing's Seattle plant on 11 December 1979.

Now the D4s are on line, the engines have made the journey from Derby, England, to Sydney, and EBQ has gone through the complex process of 'retrofit'. New engines have teamed up with aircraft in the Jet Base hangar. Back-up spare parts are available. Evaluation engineers are ready to monitor performance. Engineers have been trained to work on the D4s, new overhaul equipment has been installed, test cells organised, spare engines are positioned along the route. Some of the necessary international certification of the engines has been obtained but EBQ must, in the meantime, use its new engines at the old B2 thrust. They save five per cent

The new Rolls-Royce RB211-524D4 engines are nine per cent more fuel efficient than the Pratt & Whitney engines.

fuel over the Pratt & Whitneys; at D4 they will save another four per cent.

VH-EBQ has its own biography and its own vital statistics. It has a weight peculiar to itself, adjusted as components are removed and added, maintenance is carried out and paint layers make it heavier, Each fan blade, for example, in each of the new engines has its own individual weight. EBQ has a regularly adjusted centre of gravity. It is capable of carrying a spare engine under its left armpit — 'fifth pod certification'. The performance and age of each of EBQ's parts are itemised and logged to monitor its condition and performance. Such attention to our human parts might guarantee eternal youth.

VH-EBQ has a number, which never changes whoever the owner, given to it by Boeing when it was conceived, when Qantas ordered a hypothetical 747 and the request went into the order book: 22145. It has a model number. The basic version of the 747 is the 747-100. Before the first aircraft had even rolled off the production line Boeing announced the 747B with an improved payload and range, the 747-200B. Each airline ordering from Boeing has an allocated number: Pan Am's, for example, is 21 so a Pan Am Boeing 747-100 is known as a 747-121. Qantas' number with Boeing is 38, the fleet are 747Bs, so a Qantas 747 is a model 238B, or a 747-338 in the extended upper deck version.

The registration number VH-EBQ reveals details of ownership but like a car's registration number it is not tied to the aircraft for life. VH indicates Australia as the country of ownership. Qantas designates the rest: E (for a Boeing 747), B (if the aircraft is a 238B) or, for example, C (if it is a Combi), followed by A through the alphabet according to date of delivery. VH-EBA was Qantas' first 747, taken delivery of on 30 July 1971, and VH-EBT was the first of the extended upper decks, taken delivery of on 14 November 1984. VH-EBQ is Qantas' 17th 747-238B.

Each airline has a unique, internationally recognised two-letter prefix for the call signs that identify its flights. Each also has a striking visual logo. Left to right: Singapore Airlines (SQ), Qantas Airways (QF), British Airways (BA), Alitalia (AZ).

AIRCRAFT

Aviation parallels maritime practice in many ways, and passenger aircraft, like ships, are nearly always given a personal name. Qantas aircraft commemorate Australian cities. EBQ honours the Western Australian city of Bunbury, with 23,000 people, named after Lieut. H. W. Bunbury, 21st Regiment, who recommended the site in 1837.

3. DEPARTURE

Controller: 'Qantas One. Hold short of runway two seven. Traffic on short final.'

From the glass cabin on top of the control tower the Boeing 747 waiting to turn into runway 27 looks like a crouching grasshopper. It is quiet behind the big windows of the tower. Permission to 'push back' — leave the parking bay — was granted to the 747 five minutes ago by the controller supervising all ground movements at Melbourne's Tullamarine Airport. Now control is with the tower, and pilot and controller talk to each other over the radio.

The pilot holds the 747 behind double white lines at the entrance to the runway. The aircraft he is waiting for, a small private propjet, lands and moves down the runway. Air traffic controllers deal with each aircraft strictly in turn. They are not concerned with airline company schedules. If a plane is running late, that cannot be their worry.

Controller: 'Qantas One. Taxi into position and hold. Runway two seven.'

Away in the distance across the grass and the air shimmer, the 747 travels slowly forward and swings around to face the length of the runway. The pilot will depart on a Standard Instrument Departure (SID), which means that take-off and climb are according to a worked-out route, with a published map which the Captain has on the flight deck. This removes the need for constant communication between controller and pilot.

Pilot: 'Qantas One cleared into position, runway two seven.'

Controller: 'Qantas One cleared for take-off.'

Fifty feet off the runway QF1 is airborne. Control passes

instantly down the tower to a dimly lit room. Here the air traffic controllers who supervise the space around the airport — a distance of about 100 kilometres (60 miles) across — sit with headphones, working to radar sets. There is no view of the runways or the sky. The big room is quiet with concentration. Backs bend towards the illuminated screens. Qantas Flight One is a green blip on a departure controller's radar screen, continually reforming as the transmitting antenna rotates, picking up pulses bounced back from the aircraft.

The controller communicates with QF1's pilot on a new frequency, confirming the aircraft's height, its direction, and which radio beacon it must fly over. QF1 flies a prearranged route, adjusted if necessary by the controller to keep all aircraft clear of each other. Melbourne air traffic control centre manages the airspace of five closely situated airports.

After nine minutes and 30 miles QF1 reaches the edge of the departure controller's airspace.

'Qantas One call Melbourne control now.'

And the controller sitting in front of the radar screen hands QF1 over to the next section of sky, controlled from Melbourne by a new controller using a different radio frequency.

The trip from Melbourne to Sydney should take 57 minutes. At 37,000 feet VH-EBQ *City of Bunbury* irons out the mountain ranges and river crossings. The flight is a kind of prelude. Less than one-third of the seats are occupied because the majority of international passengers board in Sydney. Seven out of the 108 economy passengers on board are medical cases, two requiring wheelchairs. The Australia–England route is heavily weighted with what the airline calls VFR — Visiting Friends and Relations. The world has split families, and parents travel to see grown-up children in distant places, one-time migrants make journeys home, young families cross the world to see grandparents. But the sectors on the route are very long. The passengers with medical conditions have told the airline of their needs, and their details are clearly stated on separate forms: assistance required with boarding or baggage; travelling alone; must have an aisle seat or be near a toilet. Phrases are carefully chosen. 'Emotional conditions well controlled by medication' can equal alcoholic.

Up on the flight deck the crew are following a kind of guide wire in the sky. Aircraft cannot fly just anywhere. The earth's atmosphere seems a big space, but air-breathing aeroplanes can use only the first 50,000 feet or so above the ground. The sky is parcelled up into highways and junctions, broad sweeps of road and narrow, tightly controlled intersections, no-go areas and free-for-all spaces. It has to be. In April 1922 two passenger aircraft on scheduled flights collided in mid-air over France because the pilot of one aeroplane was leaning out over the edge of his cockpit, navigating by watching the line of a road down below. So was the pilot of the other aeroplane, travelling above the same road but from the opposite direction. The sky is three dimensional – up and down, forward and backward, and sideways. The pilot must know where he is in that space. And he must have rules to avoid colliding with other aircraft. The rules of aerial navigation, and air traffic control, have developed together.

The first air traffic control began at an airport near Newark, New Jersey, in 1935, with one blackboard and a map table. Today a 747 carrying passengers from one side of the world to the other is surrounded by control, one section divided from the next, with hand-over procedures at each boundary.

Passenger aircraft fly through the sky on marked-out routes called 'airways', named and colour-coded on special maps. These routes are corridors or tubes of airspace, following radio beams

Ground controllers in the control tower maintain visual contact with the ground and monitor all ground movements of aircraft and vehicles at an airport.

which radiate out, like spokes on a wheel, from ground radio stations, which are the route junctions on the airways map. Dials on the flight deck tell the pilot how far he is from a radio station, and in what direction that station is. The radio stations are the means by which air traffic controllers and pilots locate an aircraft at any time. Instructions are given and received with reference to the known bearings, in latitude and longitude, of each station. The flight paths are also marked by reporting points, quite separate from the radio stations, with invented names of five letters. The exact latitude and longitude of each is known.

The sky is divided into layers for safer flying, one layer on top of the last. The layers, called 'flight levels', are measured in feet the world over. Up to 29,000 feet — the height of Mount Everest — each layer is 1000 feet thick. Above 29,000 feet the layers are in 2000-foot stages. An aircraft flying along an airway corridor must fly within a layer. This gives vertical distance from other aircraft — what is called 'separation'. There are internationally agreed rules of separation. Other aircraft must also, in general, be 10 minutes' flying time in front or behind, and 10 miles on either side, of any aircraft flying along an airway corridor.

At the Air Traffic Control Centre, controllers use radar screens to supervise the airspace around an airport and control the arrival and departure of aircraft.

Air traffic controllers monitor aircraft in three ways. An aircraft is visible as a blip on a radar screen. Most scheduled commercial aircraft transmit a signal with an instrument called a 'transponder', which enables each one to be identified as a particular aircraft. The blip is then not just a large jet but a Qantas Boeing 747, flight QF1, on course to Bahrain. An aircraft is also identified and controlled by radio contact. And air traffic controllers identify it by a flight progress strip — a narrow piece of wood or metal holding a strip of paper marked with flight data and progress. Within an airport control area the flight strip is passed from controller to controller — physical proof of the aircraft's presence above, in the skies. The flight progress strip is a legal requirement. There must be only one strip per aircraft, to avoid double handling with its risk of conflicting instructions.

Once in the air an aeroplane cannot stop or go into reverse, like other vehicles. A big jet like the 747 cannot fly slowly. It has to keep speeding along. Approaching an airport to land, for example, the 747 must travel from two and a half to four miles every minute. So decisions about where and when to land must be made rapidly and accurately.

The air traffic controllers' job is to keep air traffic moving efficiently and quickly, and avoid collisions. They create a 'collision avoidance box' around each aircraft in three dimensions — length, height and width. This imaginary box, in which the aircraft moves on the ground and through the air, is built of real measurements. It is built of distances. An aircraft can fly higher or lower, more to the left or right. But a pilot flying within air traffic control must ask permission before he makes a move in any direction beyond certain limits.

So the journey from Melbourne to Sydney goes by, taking 19 minutes to the first navigation point, a distance of 117 nautical miles, but only seven minutes for the next 60 miles because the ground speed increases to 500 knots. The next navigational point, situated 90 miles further on, is reached in 11 minutes at a ground speed of 505 knots and is the point at which EBQ begins its descent. After 20 more minutes QF1 touches down at Sydney's Kingsford Smith Airport. It is 1.55 in the afternoon, 13.55 hours on the 24-hour clock, and 390 nautical miles of the journey have been travelled.

The landing is a little bumpy. 'Heavy-footed,' says the Captain, and apologises for a 'directional control problem', using the technical phrases of the technically minded.

4. ANATOMY OF A 747

EBQ is parked 'airside', at right angles to the terminal building. Only authorised personnel are allowed here. Doors, gates or checkpoints prevent the public from encroaching. The ground is a working surface, stained concrete set with fuel outlets and power connecting points. Specialised vehicles use this space, dominating any humans.

EBQ's gleaming silver belly is held off the ground by four sets of four-wheel bogies, like a giant's roller skates, plus a dual-wheel nose gear. The tubeless tyres hardly seem big enough for the size of the aeroplane but they have 32-ply walls — 32 layers of vulcanised rubber and fabric — compared with two or four on the family car. They are filled with nitrogen to minimise the risk of a tyre burst. Aeroplane tyres have to cope with tremendous heat and pressure increases. At take-off the 747 pivots momentarily at its centre of gravity, around door 3, leaning its massive weight on the eight rear tyres. Seconds after becoming airborne, the nose wheels retract into their wheel well, spinning tyres hitting brake pads at the top of the cavity. If the eight wing-mounted undercarriage tyres were still spinning as they were pulled inwards up into their fuselage space, the gyroscopic effect of the wheels spinning could result in damage to them, or to the undercarriage. So all fuselage wheels are automatically braked before they retract.

Underneath EBQ, the curve of the fuselage flares on and back, a running elegant line tapering up towards the nose and tail, a hand's reach above head-height. The fuselage skin is made of lightweight Alclad, a filling of aluminium/copper alloy between an outer and inner layer of corrosion-resisting aluminium. Matchstick

Location of fuel tanks on the Boeing 747.

When the fuel tanks are full, the tips of the aircraft's wings drop one metre and the fuselage sinks about 15 centimetres over the landing gear.

thick, though some parts are thinner than others. Pared down to save weight, but functioning like an eggshell, fundamentally strong and strengthened on the inside by frames and spars, and Boeing-designed tear-stoppers.

The great triangle of the wing is swept back at an angle of 37½ degrees from the fuselage. Comparisons with the size of basketball courts and ground plans of houses pale against the reality of the bulk, the sense of strength. The engine pods seem to hang, thrust forward on slender stalks. Yet a man can stand upright with arms raised inside the diameter of the air intake, and each Rolls-Royce engine weighs seven and a half tonnes. Each engine is attached to its stalk by eight expensive, precision-engineered, complex bolts. When the Rolls-Royce RB211-524D4s were fitted, one bolt cost $A150 and one engine $A4 million, making EBQ worth around $A80 million.

The wings are the flight of the aeroplane, aerodynamically smooth but able to grow and sprout surfaces to control take-off and landing. The wings are the strength of the aeroplane, carrying the weight of the fuselage and the engines in flight. The wings are not solid, rigid structures. They are hollow and flexible, run through with electrical and hydraulic systems, able to bend up to several metres at the tips. And they are the aircraft's fuel tanks. Boeing 747s are 'wet wing' aircraft, the wings carrying much of the fuel. The wing skins are milled out of solid slabs of aluminium alloy, cavities and channels etched into the inner side, weight gouged out with computer accuracy, leaving an interior surface corrugated or 'webbed' for strength. The seamless slabs are riveted together creating a great wing box, two metres (seven feet) thick in the middle. The front wing spar, the rear spar and the top and bottom skin form the boundaries to three fuel tanks per wing, with a surge or vent tank at the tip. A central tank fills the width of the fuselage where the wings join it. Huge, cavernous, the size of a large room but only stooping height, and divided into sections to stop the fuel slopping around.

The inside of each fuel tank has been sealed to create a leakproof fuel-tight container. Men with pressure guns pushing out a special sealant have covered every metal cap over every rivet. A fungus which eats aluminium can grow in the dark world of a fuel tank. It grows only in the interface between water and kerosene.

Under EBQ's passenger deck is a chopped-off curve like the hold of a ship. Soft, lightweight insulation blankets, dull silver like the lining of old tea-chests, cover the inside skin of the aircraft, wrap around the walls of the wheel wells, drape over the spars and frames. A space in the nose of the aeroplane, behind the pressure bulkhead and the cone of the radar, is like a small, empty cave.

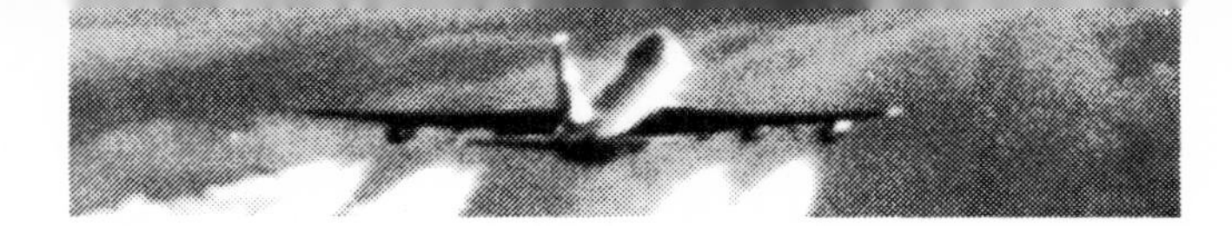

Interior metal surfaces in these under-floor spaces are painted grey-white, or apple-green — Boeing green — a zinc chromate primer which isn't damaged by hydraulic fluid. Two white metal trunks are stored by the nose wheel wall, containing Qantas spares, or parts difficult to procure along the route.

All the landing gear fits up into this belly of the aircraft, along with the centre wing fuel tank, four cargo holds and the lower lobe galley. There is a small, narrow room, painted white, warm, habitable and lagged with silver wadding: the main electronics centre, where the 'brains' behind the dials on the flight deck are kept. Black metal boxes are stored on open shelves, with space for air to circulate and keep the electronics at the correct temperature. Three Inertial Navigation System computers, the size of long filing boxes, are stored near the floor, the tops made of mesh to keep them cool. Wires in long ordered bundles trace journeys — fine-gauge, white-covered wires, wires with diagonal stripes, thick cables. Thousands of wires depart and arrive from this nerve centre.

A separate small electronics centre is stowed in several white boxes under the floor in cargo hold 1 and 2 — a big room with an entrance the width of a garage door. Hydraulic and electrical systems for the aircraft run below the floor and between the ceiling and cabin floor above. Emergency-use oxygen canisters are stored along the curve of the walls. Three large, lagged tanks holding potable water are lined up at the back, in front of the centre wing fuel tank. Two metal spheres, like large ball cocks, contain an inert gas ready to be released into the holds should fire occur. The waste water drain passing under the floor is lagged against freezing. The 10-centimetre (four-inch) diameter toilet drain coming down the side wall does not need to be lagged because it is only used when the tanks upstairs are drained at airport stops. If there were a leak in a tank, the contents could travel down the drainpipe and could freeze. But the contents cannot fall out of the aircraft in flight. The pipe outlets are securely locked and covered by access doors in the fuselage skin. Frozen lumps of urine cannot hurtle down to earth. But the toilet drains, like drains in houses, can become internally blocked. Wine bottles and perfume bottles have been found down aircraft toilets, and a Qantas First Class toilet drain was once blocked by a dozen oyster shells wrapped in a hand towel.

The used water from hand basins and galley sinks does flow down the lagged waste water drains and out through three drain holes in the aircraft's belly, into the sky. The narrow-mouthed drain holes are kept warm during flight to prevent them freezing

The nose wheel

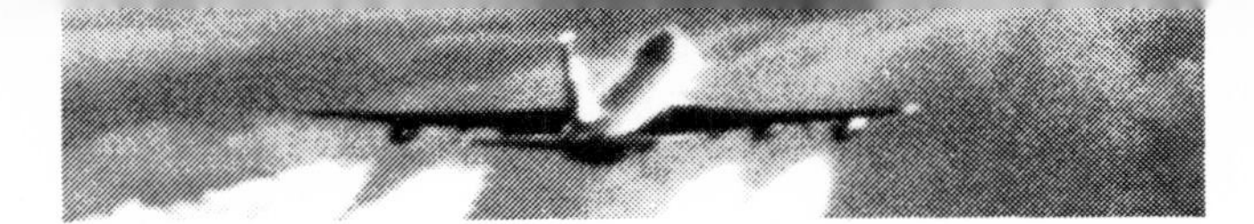

over. Water dribbles out into instant icy spray, to melt into fine mist at lower altitudes.

Behind cargo hold 5 at the back of the aircraft is a large, grubby space, low because the floor curves up here sharply towards the tail. Big pipes cross it, hydraulic systems pass along it to the rear control surfaces, and there are two holes leading outside and covered in chicken wire. Used-up pressurised air from the cabins moves into this space during flight to be expelled, breathed out, through the two gaps, the size of the openings depending on the altitude. Like the lungs of a smoker, surfaces are stained with nicotine.

EBQ's fifth engine is hidden from sight, encased behind a fireproof steel bulkhead at the very rear of the fuselage. Only the large opening of the air intake is visible. This Auxiliary Power Unit, the APU, is a small jet engine run on fuel from number 2 tank. It generates 2000 hp and has approximately one-twelfth the power of one of EBQ's Rolls-Royce D4 engines. It is a power source while the aircraft is on the ground, allowing the electrical systems to function and supplying air to the air-conditioning system.

5. COUNTDOWN TO TAKE-OFF

VH-EBQ City of Bunbury, *standing at parking bay 6 at 2.05 in the afternoon, is a container. It must be filled, over the next 60 minutes, with thousands of separate items, from 30 First Class reading books to a 79-year-old widow who has never been on an aeroplane in her life, from one packet of teething* rusks to over 80,000 kilograms of high-grade aviation fuel, from 290 Australian apples, bananas and oranges to be eaten by Economy Class passengers in 23 hours to 3295 kilograms of chilled lamb to be eaten by the Saudis, after unloading in Bahrain, and 398 bags of mail for distribution in the United Kingdom and northern Europe. Filling this container is a tightly organised event. The areas of responsibility have been carefully defined. Each dovetails into another. EBQ is the centre of a large industrious web, the outside edges of which were woven months ago, the middle sections as the components of this particular flight came together. Now, at scheduled departure time minus 70 minutes, the web is weaving tighter, to a faster tempo. As each minute ticks off and the time to departure shortens, the responsibilities for getting EBQ away on schedule increase. By 3.15 in the afternoon, 15.15 hours on the 24-hour clock, at the latest, the last threads in the web should be in place. All the timings relate to this departure time; all activities involving QF1 are a countdown to this zero.

QF1 is scheduled to depart daily from Sydney at 15.15 hours and every day the same web draws together, with the same tensions. QF29 departed Sydney for Jakarta at 13.15 hours from the centre of just such a web, and QF3 will leave for Honolulu and

San Francisco at 20.00 hours from the centre of another. And during the day all aircraft using this airport and every other airport have done or will do the same. Only the precise statement of objectives, the careful itemising of every action, the practised co-ordination of hundreds of people doing jobs they know and understand, keeps the momentum of an airport functioning.

Sydney is Qantas' home base. One side of the airport has expanded into Jet Base, an aviation town behind a security fence. Here, amongst pleasant buildings and flowering trees, technical and cabin crew are trained, flights scheduled and kept operational, and the fleet maintained. There are jet engine test cells, workshops, stores, hangars, the Flight Catering Centre, the Cargo Terminal, simulators, the pool for training crew in aircraft ditching, and hundreds of VDUs and computer terminals.

The most significant part of an airline's aircraft does not show. The outsider cannot judge how well an aircraft is being serviced and maintained. A 747 goes through a process of adjustment, modification and refurbishment. Its most fundamental or hidden parts can be replaced, from the smallest to the largest, from the vital to the cosmetic. After 10 years of service there might

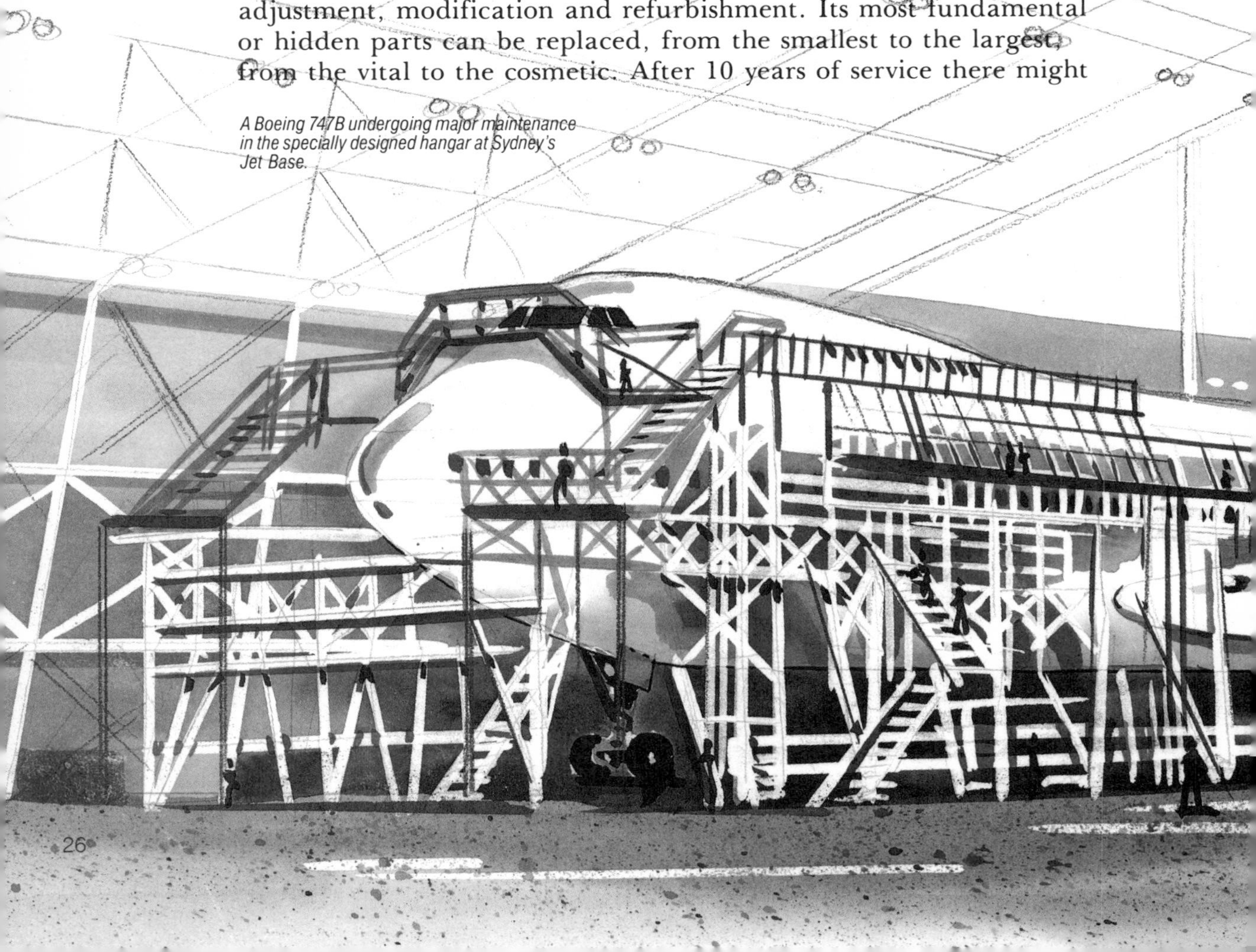

A Boeing 747B undergoing major maintenance in the specially designed hangar at Sydney's Jet Base.

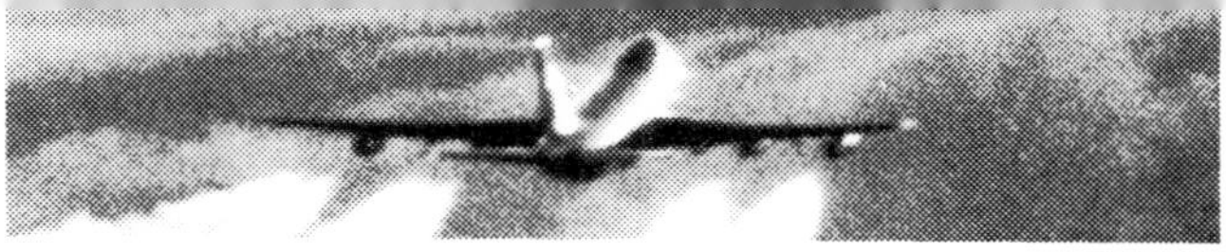

be only a very small percentage of original components still left on an aircraft. A 747 is a vastly expensive and complex piece of machinery, wearing out unevenly. Bits are being used, rather like a human's, at different rates. But whereas inadequate human teeth can only be replaced by artificial versions, the 747 receives the equivalent of new teeth, start-again eyes, young muscles, a heart at its strongest and most efficient, and supple skin. Ageing is continually monitored: worn parts are replaced and the faulty eliminated. Australia is geographically isolated and Qantas has had to develop self-sufficiency for the maintenance of its aeroplanes and the supply of spare parts. But maintenance is a complicated business, needing a large number of skills and facilities. The hydraulic and electrical systems of the aircraft — its muscles and nerves — must be able to do the jobs required of them on command. Fire extinguishers, pressure packs and oxygen bottles planted around the aircraft for emergencies must be regularly checked. The tyres, the flaps, the engines, the avionics, the wiring, the airframe skin and its painted surface all require specialist work. Sophisticated X-ray techniques, for example, are used to detect incipient metal stress in inaccessible places.

CLEARED FOR TAKE-OFF

Routine airframe servicing using a scissors truck

Maintenance programmes at Jet Base guide each aircraft through the ageing and rejuvenating process. Computers track the record of individual parts. But, quite separate from these processes, normal servicing is carried out on every aircraft on the routes, at each stop. Ground engineers do set maintenance procedures, and eight Qantas engineers have been working on EBQ since it arrived on stand.

Vehicles approach EBQ with loads, Sydney drinking water is pumped into the water tanks, fuel is pumped up into the wings, all to a precise schedule co-ordinated by a man in a yellow shirt with an office of paperwork behind him and a small piece of paper in his hand — the flight despatcher. He supervises the loading. Everything needed to get EBQ off the ground must be done on time. Any problems to do with filling the container EBQ are his to deal with.

Flights come in four types: originating — and EBQ originated in Melbourne late this morning; transit — Sydney, Singapore and Bahrain are transit stops, with the tightest timings; turnaround — which will happen in London tomorrow night; and terminating — when EBQ arrives back in Australia next Monday morning.

While EBQ is still inbound over the Blue Mountains of New South Wales, the technical crew who will fly QF1 on to Singapore are

Engineers working on a Rolls-Royce RB211 engine.

reporting to Flight Operations Headquarters, Jet Base. One Captain, aged 48, four gold stripes on his sleeves, one First Officer, aged 44, three gold stripes, and one Second Officer, aged 32, two gold stripes, report for duty at 13.45 hours. They have been working together as a team now for eight days, flying over to New Zealand and back, then up to Singapore and back, with three days off and one available if needed. The route, Sydney–London return, works out at 43 hours 40 minutes' flying time ('stick hours'), but with duty hours and other negotiated times added on, 58 hours 16 minutes are credited. Including the agreed rest periods, the trip from Sydney to London and back to Sydney will take this crew 11 days, and the five days following are completely free. After that this crew has trips to Fiji and back, a return trip to the USA and six days flying around the Tasman Sea — two months of working as a team before being rejigged into new teams.

Technical crew bid in advance for patterns of flying organised into packages of eight weeks. The most senior have first pick. In this bid period there were 102 combinations to choose from, and the crew knew which 'bid line' they had been allocated a week before the packages started. Four separate crews and four different bid lines are involved in EBQ's flight as Qantas One, Melbourne through to London, and four more on the return journey — eight complex crew schedules intermeshing, eight lots of changing seats as crew leave one flight, rest and take up another. Planning crews for flights is an art, a juggling of skills, training requirements and availability.

At the counter in the Qantas Flight Operations room each member of the tech. crew has to sign a legally binding document, a 'compliance sheet', which states that the person signing is entitled to fly the aeroplane along the routes required, is up to date on licences, and has a valid passport.

EBQ's journey as QF1, Sydney to Singapore, has been mapped out on a flight plan which is stored in a computer and sent out in advance to all air traffic controllers along the route. It states bare facts. QF1 will be flown by a Boeing 747B from Sydney, is expected to be airborne at a given time, planned to reach a certain altitude at a given place, to fly a given route and be over the boundaries of each air traffic control section by such-and-such a time. The flight plan is a legal requirement, and must be filed with air traffic control.

A 'Fuel Flight Plan' is also calculated for EBQ flying QF1 this afternoon, as the crew's work sheet. A computer takes all known data about the flight — the aircraft's weight, its payload of passengers, baggage and cargo, the speed and direction of winds expected along the route, the temperature forecast, the kind of delays possible — and works out five routes. The computer selects the shortest route which will use the least fuel as EBQ's Fuel Flight Plan. A print-out gives all details of winds and temperatures, the navigation points over which the aircraft will fly, the distances between them, the estimated time to fly those distances, the estimated ground speeds and the amount of fuel necessary for each section.

The Captain makes his own calculations about the amount of fuel he thinks he needs, based on his knowledge of the route. EBQ may have to fly at a lower altitude than planned, which would burn up more fuel, headwinds might delay it, hold-ups might use more fuel than the 1000 kilograms allowed between start-up and take-off. He adds more fuel. Most Captains do. It is 'something for Mother'.

In a back room there are two tables loaded with files of airline information, stop-press bulletins about the state of airports, engineering details and changes in the operations manual. While in here the Captain can carry out a 'safety quiz' if he wants to, a spot check on the other two pilots about the location of equipment on board the aircraft.

The Flight Engineer, with two gold stripes, 10 years' experience on the flight deck and 13 years' training before that as a ground engineer, reports in half an hour ahead of the pilots, at

13.15 hours. Flight engineers work to a different schedule from the three other technical crew. They move in and out of pilot teams because their packages are designed with a separate sequence of flights. In theory they shake up crews which might become too established together, even in an eight-week period.

The flight deck of a parked 747, with a view of the terminal wall out of the front windscreen window, is a workaday-looking place. The Flight Engineer's panel fills the height of the right-hand side, behind the First Officer's position. He sits at a narrow desk, facing the knobs, dials and switches, with a smaller panel to his right, and more controls on the rear of the ceiling panel.

The Flight Engineer checks instruments on the 747B's flight deck before take-off.

The Captain (left) and the First Officer at the controls on the flight deck of the Boeing 747B.

Zero minus 70 minutes and he checks the cockpit area — hatch down, windscreens clean, oxygen on, safety equipment in order. A detailed check is made of the systems which are his responsibility until arrival in Singapore tonight. All circuit-breakers must be in, gauges indicating engine condition must read as expected, fire systems must be checked. EBQ's technical log, its medical history, is waiting on the desk. Coupons in the log book record any 'snags' noticed by the previous Flight Engineer. These are minor problems needing to be checked by the ground engineer during this transit stop. The action taken, and a report on each snag, must be recorded in the log.

The distance from Jet Base to the aeroplane is about three kilometres (two miles). At around zero minus 70 minutes the three pilots travel across from Jet Base, 'airside' to unload their baggage, then 'landside' so the Second Officer can go through security and immigration, and on to the flight deck to begin pre-flight checks. The Captain and First Officer go for a last minute briefing to the Department of Aviation office. Then, loaded with a great weight of documentation, all legally required to be carried on board, they pass through departure controls and on to EBQ. It is zero minus 45 minutes.

The Flight Engineer is very familiar with the size of this massive aircraft. He walks around it before every flight, physically examines the carcass, shines a torch on to the back of the engines, and may even climb a stepladder to check that certain things have been done. The pilots don't often have a full view of the aeroplane. Their responsibility for flying this enormously expensive piece of equipment loaded with human beings is heavy. Perhaps it is better not to view the size of the heaviness too often.

As he arrives on the flight deck, before he takes his jacket off or puts anything down, the First Officer reaches forward and turns on the inertial navigation system (INS). Long-range jet airliners used to carry a navigator as part of the crew. When designing the new 747 in the late 1960s the Boeing Company decided not to use a human navigator. The decision was a gamble because no-one knew if aviation authorities, or pilots, would allow the huge jet aircraft to fly long distances without a skilled human navigator, and Boeing did not know if an accurate and reliable enough navigating system could be developed. But extremely sophisticated gyroscopes, and computers small enough to instal in aircraft, were available.

Adapting technologies developed for the Apollo moon flights and for nuclear submarines, a compact and delicate inertial navigation system was created for the 747 cockpit. The non-human navigator was accepted and a real break made with previous flying techniques. Navigation had taken much training, knowledge and energy. Now three INS were fitted on to the flight deck, all within reach of either control position. The human mind and body was replaced by a fist-sized keypad, a small display unit and a file-sized box.

The INS computers need a countdown time of a minimum of 16 minutes before they can operate. Equipped with its INS, EBQ will need no contact with the outside world after take-off. The INS computers do not need to refer to any outside source of information on or off the earth — no radio signals, no satellites, no lasers. They refer only to the force of gravity and the accelerations of the aircraft. The inertial system 'knows' where it is all the time because it continuously computes where the aircraft has travelled, from the place from which it took off. Each parking bay at every airport terminal of the world's airports has its exact latitude and longitude worked out and published in tables. As his initial item of aircraft preparation, the First Officer now enters this accurate position into each computer. From now on all accelerations, however small, are measured by the INS gyroscopes rotating at incredible speed, and are turned by the computers into actual direction, speed and distance travelled.

The route EBQ will fly has been defined by en route checkpoints — 'waypoints' — which are the points, defined by latitude and longitude, over which it has been planned that EBQ should fly. Up to nine waypoints can be typed into and stored in the computer at a time. The First Officer punches in the points; the Captain will check that the present position and the waypoints are accurately entered.

Every runway at every airport has its own specifications, published in a thick manual. EBQ's take-off speeds will depend on the runway to be used, its length, its height above sea-level, the outside air temperature at the time, the direction and speed of winds. The Second Officer uses published tables to calculate the speeds for EBQ's take-off this afternoon, and the speeds for the very early stages of the flight. Velocity one, V_1 — 'decision speed' — is the speed up to which it would be possible to pull up and reject the take-off. After this, any problem must be carried into the air. The aircraft takes off at a speed V_R — 'rotate' — when the nose is firmly then gently pulled up as the aircraft comes unstuck and begins to leave the ground. V_2 is a safe flying speed, given the best angle of climb for the weight, even given the loss of an engine.

Enough fuel can be pumped on board a 747 to fill the petrol tank of the family car 5000 times. EBQ will carry two-thirds of this

The fuel truck connects the fuel tanks to a hydrant in the parking bay and pumps fuel through an underground pipeline leading from the "fuel farm".

amount of fuel in its tanks for the journey from Sydney to Singapore, but not petrol. Aeroplanes fly on a high-grade, carefully purified, increasingly expensive form of kerosene — a clear, colourless liquid. Hydrants positioned in each parking bay connect to a bulk fuel reserve — a 'fuel farm' — on the edge of the airport. The vehicle which drives in under EBQ's port wing as soon as it arrives on stand is a metered fuel dispenser. Engineers connect flexible pipes up to the underwing fuel sockets and another down to the hydrant, and pumps in the fuel farm start sending the kerosene into the tanks. The connections are made with care and good design, to prevent dirt and other impurities entering the tanks. Some water is already in the tanks due to condensation. After fuelling is complete a ground engineer must bleed off the water which has settled at the bottom of each tank until a sample shows no impurities. Sometimes only a dribble of water comes out of a tank, sometimes many litres.

The Fuel Flight Plan shows that 106,000 kilograms of fuel are wanted for the journey to Singapore. The amount of fuel required is given by weight because weight is a constant. But it must be converted to volume in order to be pumped up into the aircraft's tanks, and the volume taken up by a litre or gallon of fuel varies with the temperature. There is also fuel left in the tanks from the Melbourne to Sydney trip, and the quantity actually required is calculated using tables giving specific gravity in a thick 'Fuel Book'.

EBQ's outside wing tanks are filled first, then the next tanks, working inwards. The huge room of the centre wing tank is left untenanted, except for a bit of fuel sloshing around in the base. Fuel in the fuselage is heaviest to lift.

Aviation fuel used to be so cheap the cost was insignificant. Hungry Concorde was designed in the days of fuel abundance and low price. But the cost of aviation fuel went up 10 times in the 10 years after the first major increase in 1972. Fuel usage rate became vitally important. Airline companies began a harsh period of sharply reduced profits or heavy losses. Engines had to be made more fuel efficient, weight on aircraft had to be reduced, routes had to be flown using less fuel. The money spent on fuel by an airline flying a Boeing 707 from Sydney through to London and back again in 1965 would in 1984 only take a 747 from Sydney to Darwin. Fuel and oil are now the second largest cost item in running an airline.

The amount of fuel each of EBQ's engines should need on this trip is known from monitoring. If the fuel usage rate is up — if performance falls off — the reasons are investigated.

By zero minus 50 minutes the Flight Engineer is downstairs walking around the outside of the plane, checking its general condition visually, making sure all transit maintenance has been done. He walks a set-piece route, starting at the nose gear, then up around the nose and on down the right-hand side — tyre condition, pitot tubes clear, wing tips, flaps, engines — checking for superficial damage. Zero minus 40 minutes and he must sign for the fuel.

The 747 fuselage is the width it is because the designers wanted to fit two large cargo containers side by side in the hold, and all measurements grew from that. Cargo space in the 747 is potential profit. Passengers are occupying the cabins. If cargo can be pushed into the cargo holds under the passengers, profitability goes up. Air freight is a fast-growing, very competitive business. Cargo is a valuable and attractive payload to an airline. It is much more manageable than passengers. It can fill nearly all the space available; humans need a great deal more space than their own volume. It does not, in general, need to be kept warm and pressurised and fed. It can be left in the cold or in the heat, and does not require good manners.

Cargo handling is a highly computerised business. Computers track the whereabouts of consignments, from sender through

warehousing to aircraft to destination, listing costs, weights, customs requirements, and special needs such as temperature control. Computers monitor the location in the world of the various-shaped, lightweight aluminium-alloy cargo containers, to prevent stockpiling at the end of a route. Cargo can be high security or perishable, alive or dead, outsized, dangerous or valuable. The Captain must be notified if he is carrying special loads or goods which come into the category of 'dangerous' — and where they are loaded in which hold. A consignment of perfume, for example, is classed as a flammable liquid. Polystyrene beads are expandable, evolving flammable vapour. Fibres — animal or vegetable, burnt, wet or damp — are forbidden on passenger or cargo aircraft. Aerosols should be carried in cabin baggage only. Nicotine is classed as a poison. All dangerous goods have United Nations categories, with required packaging. But once stowed on board, the Captain is the cargo's conveyor and guardian only. It is not his responsibility.

At zero minus 35 minutes EBQ's cargo is being loaded into the holds — carpet samples and butter samples, videotapes, an air compressor, news film, newspapers, chilled meat, sample motor components, a consignment of sleeping bags, another of sports shirts, and one rock pick weighing five kilograms. Uniforms, oxygen, cabin supplies and aircraft spares for the company. Cargo for Singapore, for Bahrain, some to go all the way through to London. All packed into containers and moved quickly into

Cargo compartments on the 747B are situated under the main passenger cabins. Cargo modules are loaded into these compartments through cargo doors on the right-hand side of the aircraft.

preplanned positions in each hold by small rubber tyres half buried in the floor panels. Roller tracks, power-drive systems and locking systems are all controlled by one man using a joystick on a small panel inside each door.

Sydney is a clearing house for mail. Most of the mail leaving Australia today, for unloading in Singapore, Bahrain and London is on board EBQ — two tonnes in weight. Qantas survived as a tiny pioneer airline in the early 1920s because it received a subsidy from the Australian government to carry mail. The England–Australia air route developed because governments believed that speedy carriage of mail was essential. Most governments subsidised airline companies to run airmail routes. The first aeroplanes capable of flying long-haul stages had payload space for mail, a few passengers and a little cargo, but people and goods would always be off-loaded for the mail bags. Payload space has multiplied and now bags from Australia Post make up 12 per cent of QF1's cargo weight. But airmail still has priority over cargo, and airmail is still an airline's most profitable cargo, ahead of First Class passengers. EBQ's mail is loaded at zero minus 30 minutes into bulk hold 5.

Meal production at the huge Catering Centre is computer driven. There is no fresh tomato and tarragon dressing on the avocado and asparagus terrine for Flight QF001, Sydney to Singapore, departing at 15.15 hours, without computer

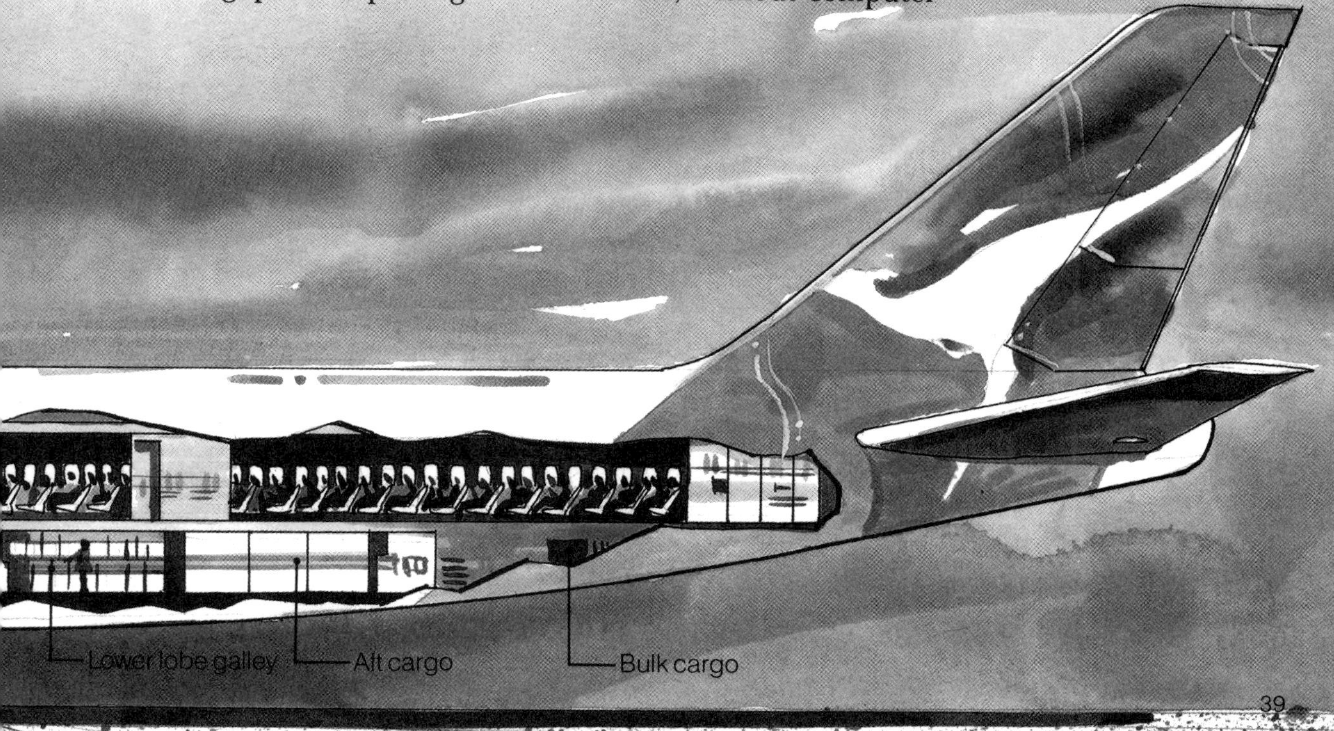

specification. No vegetarian meals for the couple sitting in seats 52F and G without computer instruction. The flight kitchens provide food for 21 international airlines as well as for Qantas. The vast river of around 1600 daily meals is channelled and subdivided, with the aid of computer print-outs, into each airline's own identifiable standard galley carts, those mobile filing cabinets for food designed to fit exactly side by side in an aircraft's galleys and to be manoeuvred along its narrow aisles. Has the price of lemons gone up? The computer adjusts the cost of each meal which uses a lemon, or even a squeeze.

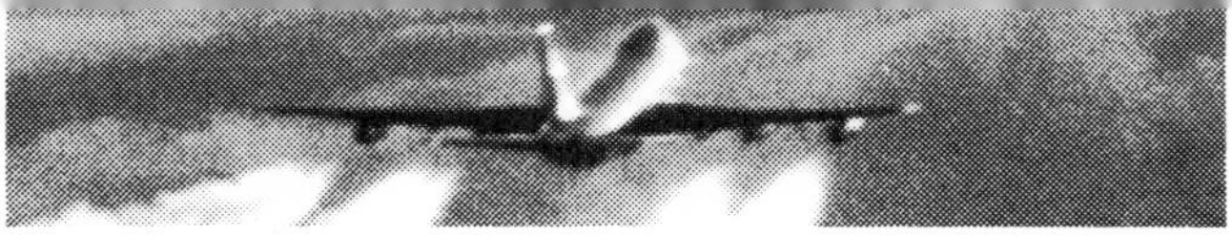

Every part of every menu for every flight is itemised and stored on the catering computer system. Seventy-two hours before a flight is due to depart details of the passenger loads are married to the menu and the raw material requirements are printed out. Forty-eight hours before take-off the production schedules go on line — how many steaks to be cooked and how many people are needed to cook them. Late bookings are added to the system. Each extra passenger means another meal.

On the day of the flight, chefs in the flight kitchens work at the mass cooking and assembly of 120 grams of duck breast and leg, 30-gram portions of carrot vichy, 40-gram portions of beans,

Trays of cold collations are prepared by chefs at the Catering Centre.

cross cut, amandine, 150 grams of fruit salad with strawberry garnish. The automatic bread-roll machine turns out 18,000 bread rolls a shift. All preparation possible is done in the kitchens. Hot meals are covered with aluminium foil. Cold meals are assembled and covered in polythene wrapping. Checkout and assembly sheets guide the meals to their correct destinations. The food is packed and stored as it will be needed in the air. Four foil packs of First Class hot savouries no. 2, for forward storage, flight QF001, Syd/Sin depart at 15.15 hours; two ditto for upper storage; 16 tossed green salads, prepacked, for cabin crew, to be stored aft; 279 Sealegs cocktails prepacked for Economy Class, stowed aft. There are 10 pages of specifications.

Before being put in an enormous refrigerator room on the edge of the catering complex, next to the loading and unloading bays, galley carts are raided by a quality checker. Meals for each class in each flight are examined. Identical-looking packages are checked for appearance and quantities.

Every aircraft must keep within a maximum take-off weight. As the figures for cargo, passengers and amount of fuel accumulate for EBQ's flight as QF1, Sydney to Singapore, their weights are calculated and fed into a computer by a load sheet

The Qantas Flight Catering Centre in Sydney is adjacent to Jet Base and the International Airport.

operator. The load sheet is updated as more information comes on line. At 14.52 hours the load sheet is finalised, a single vital sheet of computer print-out. The load controller's responsibility is to hand over to the Captain an aeroplane which is loaded within certain weight limits, in which loads are distributed to allow a safe balance, a workable centre of gravity.

Some of the weights are exactly measured; some are agreed standards. One adult with cabin baggage weighs for load purposes an arbitrary 77 kilograms. One child two to 12 years of age weighs 40 kilograms, one infant 10 kilograms. Each piece of a passenger's baggage for the hold is assumed to weigh 15 kilograms. EBQ, flying the Australia–England route in 'Pacific' configuration (28 First Class passengers, 42 Business Class and 332 Economy) has 20 crew and 400 kilograms of crew baggage. All amenities and catering needed for the route are lumped together into an agreed weight. But cargo and fuel are measured exactly; the precise number of each age range of passengers is needed, and the exact number of pieces of baggage.

EBQ has 21,449 kilograms of cargo and passengers' baggage in the holds, and 23,674 kilograms of passengers and their coats, books, flight bags, cameras, etcetera, in the cabins upstairs. EBQ has an 'empty' weight, a dry operating weight, of 179,678 kilograms, which means aeroplane plus the agreed standard weights for crew and their baggage, catering, and all amenities.

These three weights added together give a total which must be below 238,800 kilograms. EBQ's spare 13,999 kilograms is the 'underload' and the weights of last minute cargo or passengers are subtracted from this. The Captain has decided on the fuel figure and 106,000 kilograms of fuel bring EBQ's take-off weight up to 330,801 kilograms, well within the maximum of 372,000. But an aircraft can take off much heavier than it can land. The load controller adds up the amount of fuel EBQ should actually use on the trip which brings the aeroplane comfortably within its maximum landing weight of 285,700 kilograms. At take-off about one-third of EBQ's weight will be fuel. Over one-half is the 'empty' weight of the aeroplane. One-seventh will be payload.

The load sheet prints out the distribution of passengers in the various sections of the aircraft, and the distribution of fuel in wing and centre tanks. EBQ's centre of gravity is known. It must stay within certain limits as fuel is burned up during the flight and the weight alters.

Qantas cabin crew can live all over Australia. But the crew who will fly QF1, Sydney to Singapore, this afternoon must collect together at the cabin crew administration block, Jet Base, by 13.45

hours, zero minus 90 minutes. Each must sign the compliance sheet stating that they are able and prepared to fly. There is time to meet in a lounge area, chat, recognise old friends amongst other crews. The members of this crew probably will not have worked together before and probably will not again.

Ten minutes is spent in the briefing room while the Flight Service Director gives information about the flight, passenger numbers, safety procedures and company regulations. It is 'company time'. Most of the crew's work is done 'up the track', away from the Qantas base in Sydney.

Enough cabin crew have to be carried by Department of Aviation law to man all exits on a 747 in an emergency. EBQ is carrying 16 cabin crew, five more than the legal minimum — 13 men and three women, average age 34, to serve the meals and service some of the needs of around 400 passengers, and to deal with emergencies in whatever form they may arise. The likelihood is small but the training is considered essential. Twice each month, without warning, on the way to sign for a flight, every cabin crew member is questioned on emergency procedures. Anyone who fails is pulled out of the flight and a replacement put in.

Each member of the cabin crew is allocated a seat for take-off which means certain duties to do with safety are now that crew member's responsibility. All the jobs to be done during the trip go with the seat position, as a package. Senior crew can choose their seat in the aircraft and they gravitate to the front — the work is less routine. The rest are given their places by the Flight Service Director.

Zero minus 70 and the cabin crew leave Jet Base, drop off baggage, pass through outward immigration and passport control, and are on board by zero minus 55. Before boarding, the Flight Service Director does a set-piece walk in the departure terminal via the load control office for the latest passenger information, and on to the 'Captain's Club' to check on VIP passengers. The other 15 cabin crew disappear into the spaces of the aeroplane. First job is to check the safety equipment which goes with the take-off seat, then check that certain listed amenities are on board. The Flight Service Director comes on to EBQ to carry out another ward round, each action and duty clearly listed in company regulations: check the panel at door left 1 — lights, intercom, music, chimes; check the panel at door right 1; on around the cabin checking that systems, movie cartridges, headsets and tapes are all present and correct; up to the flight deck to meet the technical crew and pass on any information; around the cabin crew work stations to check that jobs are complete. It is a little like preparing to entertain. Everything is clean and tidy and the guests have not yet arrived.

6. BOARDING

The afternoon is sunny — outside the airport building windows. But QF1 passengers in transit have already officially left Australia. They are in a kind of terrestrial limbo, confined without being constrained to certain wide passages and rooms and stairs. They were processed by immigration, security and customs in Melbourne and departed Australia there. Now they are still in Australia physically, but not legally. This ordinary-looking part of a building, with taped music playing in the background, is 'sterile', nowhere. However, the passengers have come from Arrivals, downstairs, to Departures, upstairs; they have passed from one area of control to another, so they have had to go through another security check.

First and Business Class passengers are invited to rest or work in the Captain's Club, a comfortable room with newspapers, and a hostess to serve coffee or drinks or sandwiches. The most important of the flight's VIPs — the brother of a European Head of State — writes at a desk. But the atmosphere is curiously unreal and relaxed. The machinery of people's lives is in neutral. The journey for these international transit passengers has already started. Nothing more needs to be done.

In the terminal the departing passengers go through the upheaval of leaving on a long journey. They queue and worry if they have done all they should have done, if they are where they should be, and when whatever it is they should be doing next should be done. First-time air travellers stand uncertainly. The experienced look nonchalant. Farewelling parties of families and friends bump and mingle.

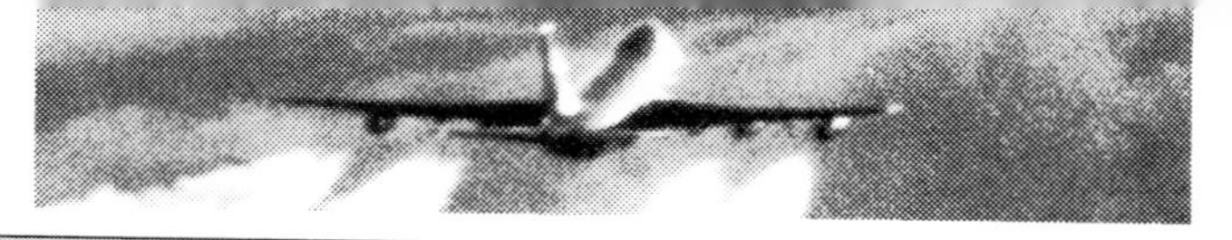

Gradually the outgoing passengers disentangle themselves from their weights and their worries — heavy luggage disappears along the conveyor belt, tickets are processed, seats allocated, boarding passes handed out. The VDU screen in front of the check-in agent shows information, already stored in the computer, about passengers. Tickets are verified against existing data. Weight in passenger aeroplanes used to be so critical that a passenger was weighed with luggage, the scales turned discreetly away from public view. The 747 is a very tolerant aeroplane and has a good weight margin. But QF1 passengers must keep within their ticket baggage allowance or pay an excess rate, and the weight of each passenger's baggage is punched into a computer for reference. Each piece is labelled with an international three-letter code for destination, and a receipt is stapled to the ticket cover. Then the baggage begins its separate journey to the aircraft, faster than its owners. Balancing on conveyor belts it travels down into the handling area. Staff read the labels and lift off each piece belonging to the flight they have been allocated to load — all QF1 to Singapore in this container; all QF1 priority to London in that. The containers used for baggage have roll-up sides and the suitcases are packed in like books in a carton, floppy bags with heavy luggage, as space permits.

Passengers cross the threshold into immigration and security; they commit themselves to the flight and pass on to the relative peace of the departure lounges where there is a curious mixture of quiet plus tension. Almost all the formalities are over now. But food, drinks, shopping and telephones are available. And the flight is yet to begin.

Airlines do not know which passengers have passed through departure formalities. Only the facts that a passenger has booked in, despatched his/her luggage and been allocated a boarding pass and a seat are on record. Each airline has one desire: to get everybody already checked on its flight through into the passenger holding area and to have the aeroplane depart on time. A certain passenger might think he knows the time, but as long as he is in the bar he is not available for boarding. Airline passenger agents with walkie-talkies patrol unobtrusively, like sheepdogs responsible for a flock at a country show. As soon as is reasonable, passengers are requested to go to the boarding lounge. Some are most anxious to be first, but there is really nothing to be first for. Seats are already allocated. At the entrance to the boarding lounge ticket slips are taken, boarding passes looked at and the numbers checked into the computer. Here at last the flock is assembling inside the final hurdles. Journeys which began this morning or yesterday, 10 kilometres or 1000 kilometres away, have been funnelled into this place. The aeroplane waits just outside the window.

Co-ordinating everything — airside activity, airport terminal areas, the state of the load sheet — is a movement office. Video screens monitor all relevant areas of the airport. Radio keeps the office in contact with the flight deck, lobby attendants and the departure gate. Zero minus 30 minutes is the critical check. Is the cabin ready to receive passengers? Are the cabin crew ready? Are the passengers tethered in the departure lounge ready to pass through the departure gate?

'Ladies and Gentlemen. Flight QF1 is ready for boarding at Gate number 6.'

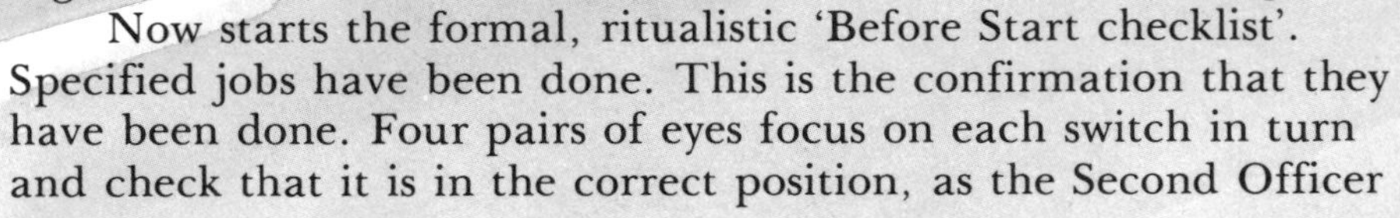

Zero minus 35 minutes and the Flight Engineer comes back to the flight deck. For the first time all four technical crew are together.

Now starts the formal, ritualistic 'Before Start checklist'. Specified jobs have been done. This is the confirmation that they have been done. Four pairs of eyes focus on each switch in turn and check that it is in the correct position, as the Second Officer

Passengers board international flights at most airports via airbridges that connect the terminal directly with the aircraft.

reads down the list. It is a litany of functions which forms part of the crew's mental and physical programming. It is a line of defence, so that, as the checklist is called out, a crew member's arm may move automatically towards a position in relation to the movement before, and so reveal that a check has been left out.

At 2.50 in the afternoon, zero minus 25 minutes, passengers begin boarding Boeing 747 *City of Bunbury* for Singapore, Bahrain and London. The anxious, the aggressive, the sophisticated, the tired, the timid, the relaxed, the excited, the already-homesick, the muddled. A glaze of uncertainty, of powerlessness, has settled over some passengers so that they fumble and shuffle. The processes they have already gone through have been too unfamiliar. Passengers go straight from building to aircraft through the enclosed tunnel of an airbridge. Several do not seem to realise that they have at last boarded the aeroplane. One plastic and carpeted interior has led to another, and now another. There is no symbolic bit of Australia to farewell, not even a last gulp of outside air.

The interior of the 747 looks huge and uncluttered as passengers come on board. It is a long, carpeted tube divided by bulkheads and curtains, the edges rounded, the colours muted yet cheerful. The rows and rows of seats sprout varieties of upholstery. Each row has a number, starting from the front of the aircraft, and each seat has a letter A through K, starting from the right as passengers face the rear of the cabin. But the positions of seats are not permanent. They can be unscrewed and arranged in different combinations. The cabin is the size of 12 average-length sitting rooms stacked one behind the other, with 14 toilets grouped at intervals, kitchens, and a staircase leading to the upper deck.

Hundreds of paying guests for a day and a night need a remarkable number of things, and quite apart from meals, 2000 items from more than 150 categories have been fitted into the aeroplane, all listed and loaded from storage shelves in the Catering Centre and brought on board at the same time as the food carts. Thirty packets of coloured pencils, 30 children's activity books, aspirin and Alka Seltzer, playing cards, stationery, disposable nappies, insecticidal spray, toothbrush and sewing kits, a clothes brush, combs, airsickness bags, cot blankets and sheets, 94 newspapers and magazines (12 copies of the *Women's Weekly* for Economy Class; *Vogue, Punch* and the *Economist* for First Class), headrest covers, pillow cases, headsets. Toilets have been equipped for passengers' needs. Spirits and wines have been counted out of the Bond Store and loaded into specially designated trolleys. The First Class liquor cart with 41 separate items, from Grand Marnier to Coca Cola, is not to be confused with Economy Class's 14 choices and throwaway styrene glasses.

One main meal and one refreshment per passenger requires an enormous range of things to serve with them and things to serve them with. There are pages and pages of stores to be counted out and loaded for each flight — oven mitts, tablecloths and cocktail napkins, soiled-linen bags, wine chilling bins, highball-glass racks, macadamia nuts, French mustard. QF1, departing Sydney, requires 332 two-thirds trays for Economy Class refreshments, 36 brandy balloons, 1600 disposable glasses, 878 toothpicks, 28 gold napkin rings and 907 Economy Class teacups.

The shelves and drawers in every galley have a specified quota of equipment. One silver caviar bowl is in the bottom shelf of number 17 cupboard, upper deck galley, but the silver cake server is in utensil drawer number 25. Knife sharpeners, icepick, pot scourers, towel tongs and dustpan are all in correct quantities and stowed in their correct places. A modified house has had to be loaded for around 400 people: their kitchen and bathroom requirements, and some of their sleeping and entertainment needs.

Qantas Operations Control Centre keeps track of every 747 in the fleet 24 hours a day and initiates action to avoid delays.

Two passengers are delayed at immigration. All departures delayed for more than three minutes, anywhere in the world, must be reported to headquarters in Sydney. The reasons are analysed at twice-weekly meetings. Delays are a matter of corporate concern. Punctuality on a long-distance flight is defined as arrival at terminal destination within 15 minutes of published schedule. Qantas aims for a long-haul arrival punctuality rate of 65 per cent, and a departure punctuality rate of 70 per cent, although actual figures achieved can go well above 80 per cent. A higher rate would be possible if schedules were padded — if enough time was allowed in each sector for delays to be made up. But schedules are competitive. Thirteen airlines, for example, currently fly the Australia–England route, and many passengers look for the fastest time.

Any delay means instant decisions. Information about the two passengers held up at immigration and their tickets is assessed. The decision whether or not to hold the aircraft is made by committee.

Then the two laggards board. Slowly and carefully, at 17 minutes past three, the great aeroplane moves backwards away from the buildings and from the airbridge which is now suspended in mid-air, leading nowhere. Departure is two minutes behind schedule — less than three minutes — so allowable as an 'on-time' departure. The runway has been changed by ground control at push-back. 'Because the northerly wind has strengthened we have a long taxi to the northern runway,' announces the Captain. The aeroplane trundles on, across the flat spaces of the airport, and waits at the entrance to the runway. The minutes are very long. Information lulls the passengers' unease at the time it all seems to be taking and the Second Officer offers explanations. A small plane lands far down the runway, slows quickly and turns off. A larger passenger aircraft lands further back, across EBQ's nose, and slows after travelling a longer distance. All delays after push-back are the responsibility of air traffic control and are out of the hands of the airline. Then, at 3.30 pm, the 747 turns into the runway, its engines roar, and the aeroplane sprints, races, and is off the ground and driving up into the air, 331 tonnes of aircraft airborne at 15.33 hours Sydney time.

At the airport the large web surrounding QF1 is being folded up and packed away. Everything has been documented in about 50 to 80 pages. A bit more documentation and then it is signed off into the flight files. 'It's not kerosene that gets aeroplanes off the ground,' says a despatcher, 'it's trees.'

QF1 is now a card on the large movement board in Qantas Operations Control Centre, where the status of each aircraft in the fleet is constantly monitored. The board across the width of the room shows the current position of all flights leaving and coming to Australia. Clocks give local time and Coordinated Universal Time (UTC). Flight control depends on efficient communications. Movement messages come in to a central computer and VDUs display updated information about each trip. Here in Operations Control, emergencies are dealt with whenever they occur. Security is co-ordinated, rerouting organised, and the implications of strikes, fuel shortages and mechanical problems are acted on. If a flight should become seriously delayed, decisions must be made quickly to minimise the repercussions.

7. IN FLIGHT

Down below, Australia's emptiness begins quickly. The Blue Mountains are crumpled-looking and dull-coloured, marked only here and there by human activity and partly hidden by a flat layer of teased-out cloud. Fleecy white 'solid' clouds wait ahead, but once the aeroplane is inside they are, disappointingly, the same as being in a grey mist.

Drinks are served. Drinking plenty of liquid helps prevent dehydration but alcohol is least useful. Business Class get cocktails and canapes. It is 3.55 pm, afternoon tea time, but already food is fitting into the number of hours the flight will take between Sydney and Singapore, irrespective of tummy time. Eating on an aeroplane does not have a great deal to do with hunger. It is an entertainment, an activity, something to look forward to and something to do. For most of the passengers the entertainment potential is limited by Economy Class tickets. First Class passengers, in their spacious surroundings, expect interesting food served well — luxury and abundance temporarily suspending diets and caution. Wedgwood bone china and silver cutlery are appropriately heavy for important eating. Aircraft weight is saved in the rear three-quarters of the aeroplane where modest meals are served in simple containers, coffee comes in melamine cups and wine in styrene throwaway glasses.

Over the corner where the borders of New South Wales and Queensland meet South Australia, 'Charlie Bravos', big cumulo-nimbus clouds, show up on the the flight deck radar screen, the information fed from the scanner in the fibreglass nose-cone. EBQ moves five miles out of track to avoid possible turbulence. All cabin service ceases while the aeroplane flies through greyness, and passengers fasten seat belts. Blanked out visibility and slight bumpiness last about seven minutes.

Around 5.00 pm the passengers in Economy Class, as if they were all members of one big family, sit down to an identical meal of seafood cocktail, beef bourguignon with noodles, passionfruit pavlova, and the kind of cheese whose packaging is entirely unwilling to release its grip. The space available would test the manners of the closest family.

It is not easy to plan a meal which will be enjoyed by most people most of the time. Eighteen passengers have officially opted out, their needs specified when they booked their tickets, and they eat their vegetarian or diabetic or chicken-only diets. Business Class passengers, like Mother Bear, have a middle-sized menu printed on medium-weight paper, and eat in their medium-sized chairs, with more leg room than Economy and less than First. They have a choice of steak or barramundi, and strawberries as well as passionfruit on their prepacked pavlova. As listed on the computer print-out, the Economy Class dinner eaten over the Simpson Desert has 12 items, Business Class scores 26, and First Class 30, plus standard extras of butter and milk, parsley sprigs, cream, plain and whipped, carnation centrepieces and a bag of lemons.

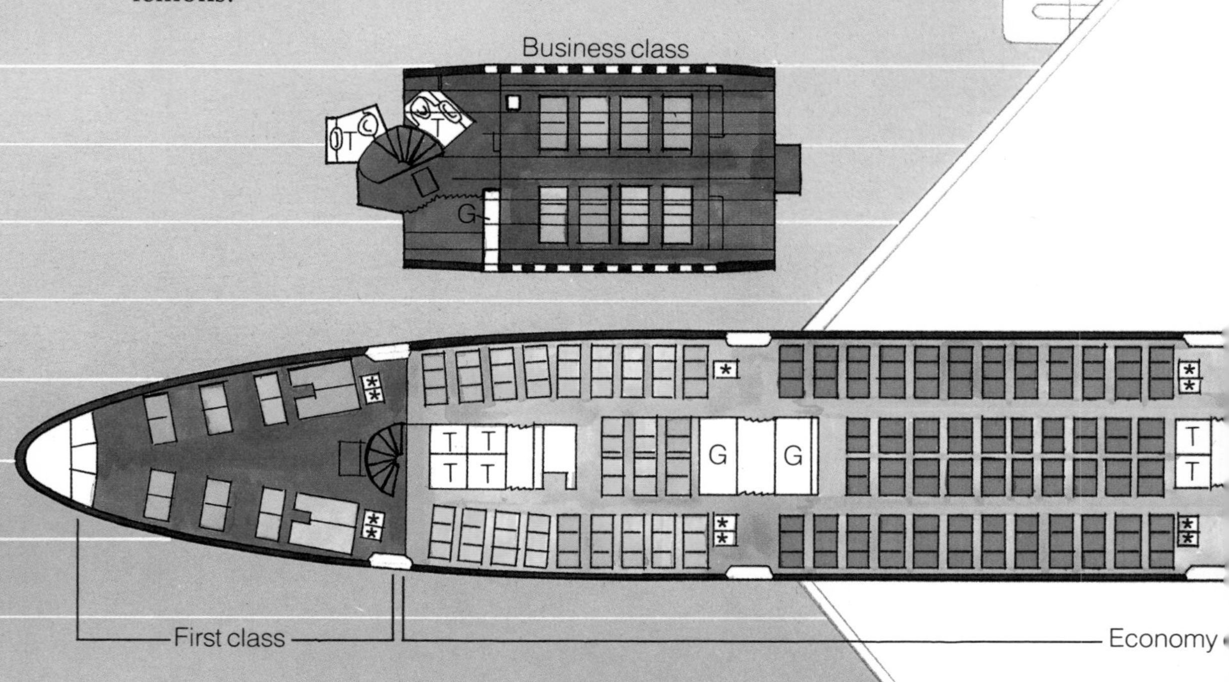

The First Class menu rivals those on the pre-war flying boats travelling the England–Australia route. On QF1, hors d'oeuvres of oysters, grapefruit and crabmeat, or avocado and asparagus terrine are offered; a choice of three main courses, including roast sirloin, and lobster in Pernod, with salad or vegetables; strawberries and sorbets, cheeses, fruit, a choice of four coffees, and Qantas handmade chocolates. On Qantas Empire Airways Short S23 Empire flying boat, travelling from Sydney to Singapore on 25 December 1938, all passengers were offered iced consommé, oysters, roast turkey, York ham, French salad, sweet Russian salad, mince pies, jelly and trifle, fruit salad, fruits, nuts and coffee.

From 32,000 feet up, the Simpson Desert is like a flat beach after a long tide has pulled back — small, hard-edged, regular ripples of sand. Charles Sturt explored the edges of this desert in 1845, searching for an inland sea. From the sky it looks as if the water has just left.

Passengers have been fed and watered and are relaxing into their accommodation. The galley carts have been repacked with the remains of meals, to be off-loaded at Singapore. The cabin crew eat their dinner in the crew rest areas — each person has 20 minutes off. There are four separate menus to share between the 16 cabin crew and four technical crew, a safety precaution so that everyone does not eat the same food. The Captain and First

Passenger seating is divided into six compartments: five on the main deck occupied by First Class and Economy Class, and one on the upper deck accommodating Business Class.

Officer must eat different meals from each other.

The Flight Service Director, as manager supervising the running of this temporary hotel, knows necessary details about some of his guests. He knows who are his VIPs and CIPs (Commercially Important Persons) and where they are seated — MPs, directors of companies, government officials, and of course the brother of the European Head of State. In seat 15C is a passenger who had a problem on a previous flight ('previously mishandled') and needs careful handling. A passenger with mild claustrophobia is seated in 21H, and one travelling under stressful circumstances in 25B. An elderly passenger in seat 47C is a first-time flyer and is travelling alone. There are now 15 medical cases on board. The Flight Service Director holds the keys to the physician's kit but it can only be used by a doctor.

The Sydney–Singapore section of the flight is the busiest for the Flight Service Director. There are papers to sort through, forms to fill out. A red satchel was brought on board in Sydney for him and it is bulging.

Five past six in the afternoon Sydney time and EBQ passes just west of Alice Springs, having reached its cruising height of 35,000 feet. The Chief Flight Attendant chats with a senior flight attendant about the race meetings this coming bank holiday weekend in England. Which will they go to — Sandown? Epsom? Ascot? A bored youth in a jacket stamped 'Australia' has wandered the length of the aeroplane. 'Where's the table tennis room?' It is getting near passenger entertainment time. One hour 50 minutes of films in the sky, while down below the sparsely textured ridges of desert Australia edge by, the smooth circles of claypans, the colours soft and intangible as daylight fades. EBQ is flying in a diagonal line across the continent, Sydney to Derby on the west coast, a route pioneered by Charles Kingsford Smith in *Southern Cross* in June 1929. The north-west coast of Australia approaches. Some people on this aeroplane are about to leave Australia for the first time. Some for the last time. But at 35,000 feet, with blinds drawn and the cinema screens showing *Night Crossing* and *Whose Life Is It Anyway*, the event is unremarkable. The departure, for passengers, took place on the east coast at the airport.

After four hours nine minutes' flying time since take-off at Sydney, EBQ crosses the coastline, having flown 1790 nautical miles, 2060 statute miles, 3315 kilometres, at a speed through the air which can be measured in three ways. 'True airspeed' (TAS) is the speed of the aircraft's passage through the air, and it increases as the aircraft climbs through the atmosphere and the pressure of the air lessens. 'Indicated airspeed' (IAS) is the speed the aircraft thinks it is passing through the air, the speed it feels itself to be doing, and the figure is essential to the control and management of the airframe in the air. 'Ground speed' is true airspeed modified by the effects of wind, so that it increases with a tailwind and decreases in a headwind. This is the speed which is of interest to passengers because it tells them how quickly they will arrive at their destination. With a seven-knot headwind, EBQ leaves Australia at a TAS of 502, an IAS of 320, and a ground speed of 495 knots. The temperature outside is minus 40°C, minus 40°F, the only temperature at which the two scales meet. Most scheduled passenger aircraft spend most of their time in this temperature band. It is 7.42 in the evening Sydney time. Twilight. If EBQ flew a little faster it would keep up with the sunset. But the speed being flown is consistent with efficient subsonic flight. Closer to the speed of sound, vibrations would begin, which would be uncomfortable for passengers and potentially damaging to this category of aeroplane, with its weight and wing design. Subsonic passenger aircraft are pegged to a maximum speed of around 85 per cent of the speed of sound. This is the state of the art.

After the great width and length of the 747, the flight deck feels small and intimate, quite separate from the bulk of the aeroplane stretching below and behind. The tech. crew are physically separated by distance, and a closed door. They are psychologically separated by their angle of vision — no view behind, only forward and to the sides. It is just possible, by craning, to glimpse a wing tip. There is a persistent, quiet background sound, a kind of soft hissing, which emphasises the lack of obtrusive noise. The windows are small, minimising the risk of a bird strike. The side windows are thick layers of stretched acrylic. The two front windscreens are thicker, laminated glass and acrylic sandwiches, each weighing 68 kilograms (150 lb). All windows must be heated for 20 minutes before the aircraft takes off, to stay flexible and resilient. A thin sheet of embedded gold layered through each window carries an electric current. The temperature of the front windows must be kept at 29°C and the side windows at 44°C during flight.

There is a subtle atmospheric difference on the flight deck compared with the cabins. Humidifiers using the precious water supply keep the cockpit area at ground humidity. The dry air downstairs causes dehydration, increasing tiredness. But the moist air is caught on the insides of the relatively uninsulated metal window frames of the flight deck and turned into hoarfrost. Just a thumb thickness away it is 40°C below freezing.

Here on the flight deck three or four men manage a cluster of complex systems. Their concern is the proper functioning of those systems and the safe, punctual delivery of the aircraft's payload. Everything is within reach — switches, dials and controls in front, above and to the sides, duplicated where necessary for each of the two pilot seats. Two small radar screens are set at calf-height, one to the left of the Captain and one to the right of the First Officer. The four throttles move forward and back on tracks in the central control pedestal. The Flight Engineer sits behind and at right angles, but his seat can power forward to just behind the central pedestal.

The crew watch the INS Control and Display Units, which can show present position, ground speed, wind direction and speed, angle of drift — displacement from the desired track, if intentional — and other navigational data. At all times the pilot knows from which waypoint the aircraft is travelling, and towards which one it is moving. He knows immediately and at any time exactly where he is — not, as with all previous navigational systems, only where he has been.

The three INS computers on Qantas 747s work in committee, pooling their results and giving the best answer. If one should start to deviate, a light comes on and the other two take over. Too wide an error and the system is removed and sent to a specially built Clean Room at Jet Base, Sydney, where temperature, humidity and cleanliness are all rigidly controlled. A gyroscope is the most delicately balanced piece of equipment in the aircraft. Maintaining it takes 300 skilled man-hours. One piece of dust in the gyroscope and the whole job of setting it accurately must begin again.

As EBQ passes over Derby, the First Officer and the Flight Engineer do a check of actual performance against forecast performance on the flight plan. EBQ has taken the most usual route, the shortest, known as Alpha 76 — Sydney–Alice Springs–Derby. Time has been made up and less fuel has been used than predicted because headwinds have not been as strong as expected. The First Officer calls up the controller at the boundary of the air traffic control zone, giving the aircraft's number and altitude — a routine safety procedure which also checks that the radio is working efficiently. If EBQ had not called within three minutes of the time that it was expected over the boundary, known from the flight plan and from the actual departure time sent along the network, the controller would have called the aircraft, using its own unique personal 'number', DKFM. No other aircraft has this SELCAL identification number and EBQ is contactable anywhere by radio via it, depending on atmospherics.

There are no aircraft ahead and one British Airways 747 is behind, flying at 31,000 feet. Commercial air traffic over Australia is one way — leaving in the afternoon and arriving in the

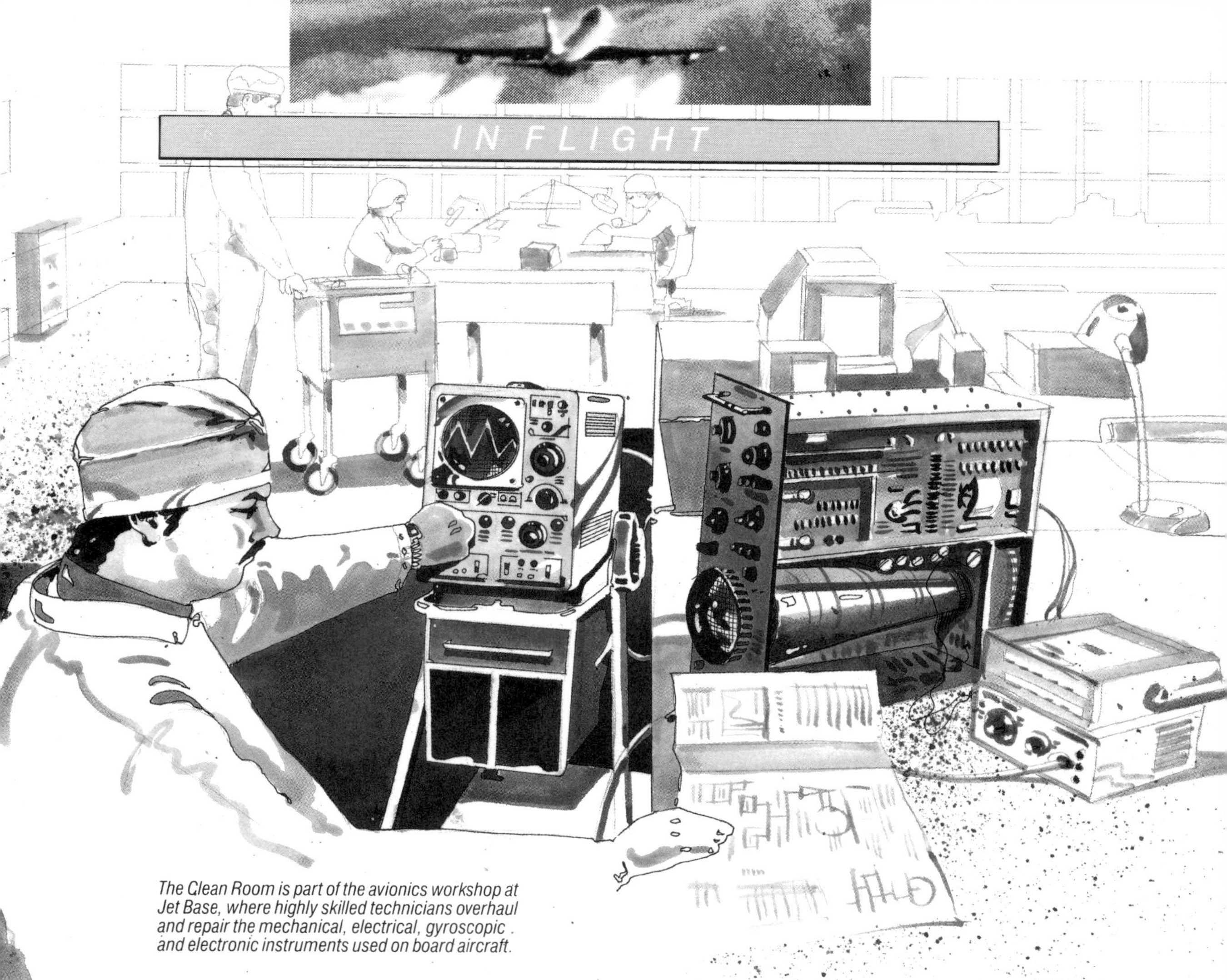

The Clean Room is part of the avionics workshop at Jet Base, where highly skilled technicians overhaul and repair the mechanical, electrical, gyroscopic and electronic instruments used on board aircraft.

morning. Pilots have more chance of flying at the altitude they want over Australia. The skies are empty compared with Europe.

One hour after passing Derby, and 720 miles south of the equator, EBQ leaves Australian air traffic control and enters Indonesian control. The films end. It is 8.40 pm Sydney time and five hours have been spent in the air. The darkness outside will last now until eastern Europe. Organised activity slows down and passengers rest. In an hour's time a half-meal will be served — 'refreshments' — sandwiches and cakes for all classes, with tea or coffee. First Class have crab, salmon or beef in their thinly sliced bread, and a tart; Economy Class have ham and cheese rolls, with cookies.

Thirty minutes out from Singapore's Changi Airport the flight gently begins to wind up. The temperature in Singapore is 27°C and it is rather humid, passengers are informed. Conditions are fine, and local time is two hours behind Sydney time.

The Captain still has his watch on Sydney time — 'K' or 'Kilo' time in aviation language. For convenience, the letters of the alphabet are used to describe bands of longitude. UTC, Coordinated Universal Time, the old Greenwich Mean Time, is called 'Zulu' time. Rome, seven and a half degrees east of Greenwich in London, is 'Alpha' time. For almost all Qantas crew,

normal time — body time — is Kilo time. But the clocks on the flight deck are set to UTC. Aviation has to operate to a single standard and all aviation business is done according to UTC. Wherever an aircraft may be in the sky or on the land, whoever is flying it or talking to it, all must work to the same time.

A heavy aeroplane like the 747 travelling at speed can use its own weight to glide. One hundred and ten miles out from Singapore EBQ begins its descent, travelling roughly three miles for every 1000 feet of descent, or 15 parts forward for every one part down. The four throttle levers are back as far as they can go and the great engines are on idle. Descent control clearance comes from the approach controller at Singapore to descend in stages to 1900 feet. The altitude winds down — 1000 feet every 20 seconds now. The controller's area of influence extends 100 kilometres (60 miles) out, 10 minutes' flying time from touchdown. Below is a small island of 2.5 million souls and four airports. Changi has been built out onto reclaimed land, its solid, level, strengthened slabs of runway ready to withstand the weight of a loaded 747 meeting the ground at 130 knots.

After almost eight hours of flying EBQ approaches runway 20R with accuracy, in the darkness. The statement and response between control tower and flight deck continues unhurriedly, calmly. At 1500 feet, on final approach, power goes back on to an approach-thrust setting. The Captain takes over from the autopilot five miles, and two minutes, from the runway.

At 11.05 in the evening Sydney time (9.05 pm Singapore time) the cabin crew check that passengers are seated upright with seat belts on and that all cigarettes are extinguished. At 11.08 pm the undercarriage is lowered with a distant clunk. The wing edges are unfolding, a metal flowering of gaps and curves, like a hand stretching its fingers. At 11.13 the cabin crew are strapped into their seats. At 11.15 there is a slight bounce, and touchdown. The engines roar in reverse thrust, and in 40 seconds EBQ is ambling along the runway at taxiing speed. At 11.20 pm Sydney time, 9.20 pm Singapore time, EBQ is docked, nose to the wall, at Parking Bay E 79. Touchdown has been made within Changi's peak hours of busyness, when one aeroplane lands every four minutes.

8. IN TRANSIT

Up on the flight deck the ritual of the 'Shut Down checklist' is being called. The engines are quiet, and the tech. crew confirm that the aircraft is secured. They finish the flight's paperwork, and fill in the deck logs which stay on board until EBQ's journey uphill to London and back downhill to Australia is over, on Monday. The inertial navigation systems have brought EBQ over land and ocean. On average they are three to four miles off course at the end of an eight-hour flight. The exact position co-ordinates of this parking bay, E 79, are checked against the three INS results, and the engineer records the error rate — five miles on 1, nine on 2 and one on 3. He notes the fuel consumption of each engine. All results will be sent back to Sydney, plus the exact time of arrival. Fuel usage and punctuality are monitored at corporate level.

The Captain's responsibility for EBQ ended when the engines shut down. The new Captain takes over when the engines start up again in about an hour. In the meantime, here at Changi, responsibility for the aeroplane is with the ground engineers. The tech. crew leave the Flight Deck, get through customs and immigration and into waiting transport for their hotel. The new tech. crew are going through the process of signing in at the Qantas Flight Operations room. There must always be five cabin crew members on the main deck of the aeroplane and one on the upper deck during a transit stop. The arriving cabin crew greet the departing. Who won the footy? What's it been like? Then the old crew are through customs and immigration and on to the waiting bus. There are 25 hours five minutes' transit time free in Singapore

but the crew must not go away — no trips into Malaysia or cruising on a yacht. They must be contactable. Tomorrow night, a Saturday, they will be back at the airport. Qantas Flight One, which will depart Melbourne on Saturday lunchtime, will fly into Changi, a tired crew will sign off, and they, refreshed by around 24 hours of 'slip time', time without duty, will take the aeroplane on to Bahrain.

EBQ has one hour five minutes' transit time in Singapore. Time to arrive, unload passengers and the correct cargo, mail and baggage, take away leftover food and used catering equipment and toilet waste, clean the aircraft, replenish it with fuel, water and food, refill the holds with cargo, mail and baggage, carry out maintenance checks, load passengers, check the systems and depart.

Transit time at an airport is worked out according to how quickly everything can be done, plus about five minutes. But how quickly everything can be done at each airport depends on what fuelling systems are available, how congested the airport becomes, how efficiently the airport is designed and laid out. As EBQ parks at Changi, Singapore Airlines equipment under contract to Qantas is lined up airside ready to pounce — to surround the inert body of the great aeroplane like attentive bees servicing the queen. The airbridge uncoils as the aircraft docks. The first catering truck drives up to the starboard door and begins unloading even before passengers have got themselves into line for disembarking. The refuelling tanker is already nuzzling under the wings and the 'honey cart' is feeling for the outlet drains of the toilets. Baggage is being unloaded at the door of cargo hold 4 and cargo pallets slide out of hold 1 onto the platform of a waiting scissors truck. Ground engineers are testing the oil. Cleaning ladies stand by with boxes of fluids and cloths, brooms and vacuum cleaners, to occupy the cabin. The cabin dressers wait to add headrests, in-flight magazines, toilet paper and soap. The action is orderly but fast. Punctuality depends on achieving specific time targets.

At Singapore, passengers used to climb out of their aircraft into muggy warmth and know they had arrived in the tropics. The aeroplanes parked out in the open like ships anchored in a harbour, and buses, like launches, ploughed across busy tarmac areas to distant airport buildings. Now passengers unload straight into air-conditioned terminal buildings and there is no smell and feel of Singapore. Changi's concern is the efficient, pleasant channelling of huge numbers of international passengers to and from their aircraft. Moving walkways save their legs. Shops save them going anywhere else to shop. There are fountains and flowers and vast clean spaces, but some QF1 passengers cluster a little

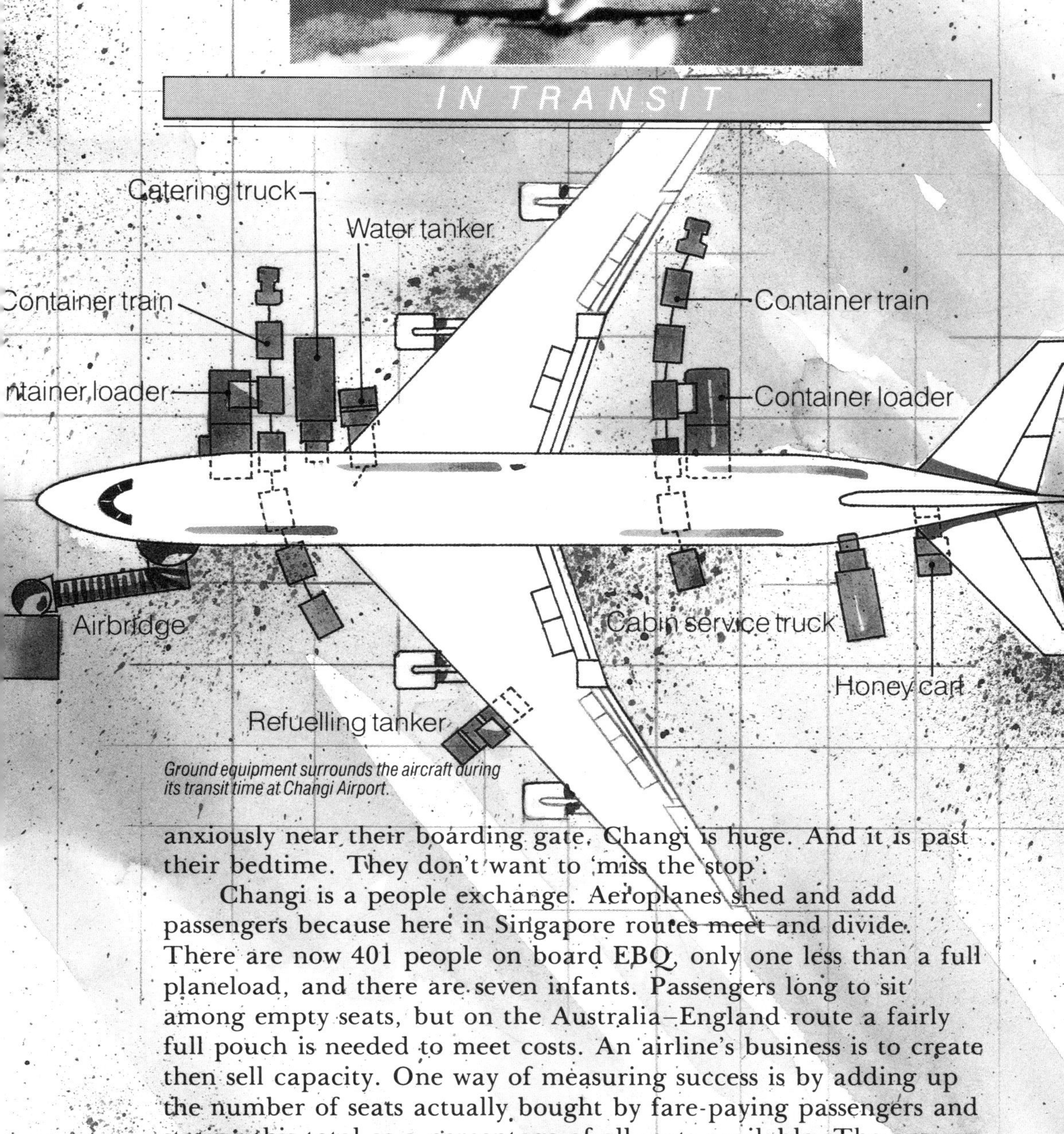

Ground equipment surrounds the aircraft during its transit time at Changi Airport.

anxiously near their boarding gate. Changi is huge. And it is past their bedtime. They don't want to 'miss the stop'.

Changi is a people exchange. Aeroplanes shed and add passengers because here in Singapore routes meet and divide. There are now 401 people on board EBQ, only one less than a full planeload, and there are seven infants. Passengers long to sit among empty seats, but on the Australia–England route a fairly full pouch is needed to meet costs. An airline's business is to create then sell capacity. One way of measuring success is by adding up the number of seats actually bought by fare-paying passengers and stating this total as a percentage of all seats available. The answer, called 'revenue seat factor', is a useful statistic, comparable with, for example, 'break-even seat factor', which is the percentage of seats which need to be filled before expenditure equals revenue. The demand for seats on an aeroplane varies greatly according to the day of the week and the season of the year. The England–Australia route suffers from directional imbalance — peaks and troughs. Planes are full going into Australia for the summer season, but not so many people want to travel to England in the middle of winter.

The revenue from selling seats varies according to the route. Seventy-five per cent of England–Australia passengers buy the cheapest fares they can, leaving 25 per cent paying First, Business or Economy. The yield from the route is lower for every passenger kilometre travelled than, for example, the route to Japan, where only 50 per cent of passengers travel on the cheapest fares. But business expense accounts pay for many trips to Japan. Most people pay for their own travel between England and Australia. The costs of operating aeroplanes on particular routes varies widely. Fuel is more expensive at some stopovers than others, and proximity to oil-producing areas is no guarantee of cheapness. Airports charge a range of landing and handling fees. On the England–Australia route, Heathrow's charges are high, and Qantas 747s must wait in London all day before departing again.

It is 10.00 pm Singapore time, midnight in Sydney, 1400 UTC. Action on the flight deck is cool but very precise. The Captain is in the left-hand seat, First Officer on the right, Second Officer behind the Captain, Flight Engineer at his bank of instruments and switches on the right-hand side. The walls of the parking bay seem to wrap around the flight deck — the umbilical cord of the airbridge still connects aircraft to terminal.

But away from the brightly lit terminal and its 24-hour activity it is night-time. Away from the air-conditioning the air is warm, damp and tropical. Changi airport is very busy. Floodlit tailplanes of huge jets gleam out of the dark. Lights in EBQ's stabiliser reveal the Qantas logo, the flying kangaroo. The logo lights will stay on through take-off and climb.

The aircraft is pushed backwards away from the parking bay by a heavy-duty tug connected to the nose wheel.

The lights of the flight deck are fully on. All four officers are working, checking systems and equipment against a list read out by the Second Officer. The exact position co-ordinates, N.01.21.5, E.103.59.6, have been fed into the INS computer and the first nine navigation points punched in. Speeds for the take-off have been worked out. The departure map (SID) is studied: out to radio beacon Sinjon and over it at 6000 feet, turn and come in over radio beacon Johor Bahru, and cross the border into Malaysia at less than 25,000 feet.

The Captain briefs the crew for the take-off — a ritualistic recital of everything they will do at every moment of the take-off and climb-out, and action in the event of any failures. The ramp co-ordinator delivers the load sheet, a computer print-out of EBQ's final weight and balance, with the exact number of passengers, their baggage weight, the cargo weight and the total fuel on board. The Captain checks and signs it.

'We are at Gate 79.' The Captain is talking to air traffic control.

'Qantas One you are cleared for flight.'

'Ready for push-back,' says the Captain. 'Brakes are released, ground.'

Off-chocks time is 10.32 pm local time, 1432 UTC. A 17-minute delay but this is within the departure slot given by air traffic control. The Captain is keen to get away fast. Altitudes are

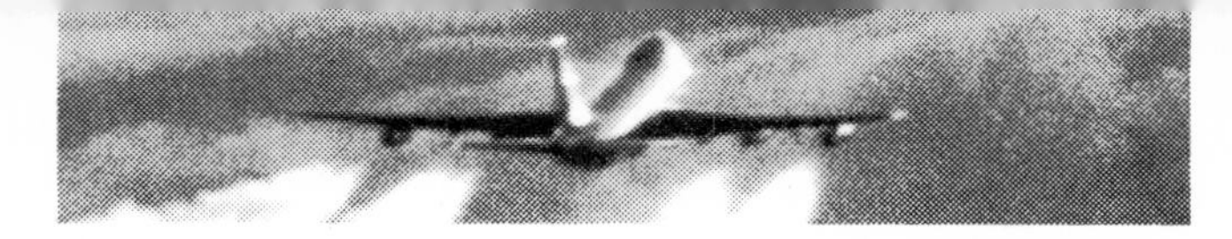

allocated according to who leaves first. All pilots believe an airport favours its own airline. The airbridge has snaked back against the terminal. The 747 moves backwards away from the parking bay, pushed by a strong, squat tug.

'And we are starting one.' Engine number one joins number four. The lights are dim on the flight deck. The instrument dials are softly lit. There is a sense of anticipation, almost exhilaration.

'Qantas One taxi please,' orders the control tower. 'Two zero right.' It is 1438 UTC. Everyone looks out of the windows, watching, watching. A map of the airport taxiways and runways is in front of the Captain.

'Qantas One take first turning left,' instructs the control tower. The Flight Engineer continues his checks on the aircraft systems, flicking switches, reporting his actions. The red-covered Flight Engineer Quick Reference Procedure Checklist is in front of him. All tech. crew must follow written procedures with exact precision.

Captain: 'And Qantas One is ready.'

The First Officer is doing the take-off. At 10.46 pm Singapore time, 1446 UTC, EBQ powers along the runway towards the south-south-west at 80 knots. At 10.47 'Rotate', and the First Officer pulls back on the yoke, the nose lifts off the ground and Qantas One is airborne.

Control tower: 'Qantas One airborne at time '47. Call Singapore Departures. Goodnight.'

Out over Sinjon four minutes later the deep-black sea below is set with the lights of anchored ships. 'At 5000 feet Qantas One,' says control. EBQ swings in towards the city, climbing away, leaving streets edged with lights stretching out between buildings like a net, holding fewer and fewer lights in each space as the aeroplane reaches the border. It happens so fast. EBQ is so low above the sea and the city — Singapore is all there, shining in the dark — and then it's gone, and the plane settles into even flight. Singapore control has handed EBQ over to Kuala Lumpur control. The autopilot is already on. The time of intense concentration is over. The crew relax and undo their shoulder restraints.

Singapore Airlines and Swissair are both ahead, so EBQ flies at 28,000 feet, lower than the crew would like. Flying at 28,000 feet uses more fuel, which everyone wants to avoid. The air is warmer and engines produce less thrust here than in the colder air of higher altitudes. Singapore Airlines has the 31,000-foot band, and Swissair, 35,000. Bands at 33,000 and 37,000 are reserved for aircraft flying into Singapore. No-one is at 39,000, but EBQ is too heavy to get that high yet. There is a full passenger load.

9. NIGHT FLYING

The cabin is warm, with 28 First Class passengers, 41 Business Class and 332 Economy Class, plus seven infants. All those bodies create their own warmth, raising the temperature. The Flight Engineer notes the rise of temperature and lowers it a bit. Being inside the 747 is like living in a shared spacesuit. The air is pressurised to around 8000 feet, kept to about 22°C and changed continuously so fresh air replaces old air. The engines use air at a formidable rate but there is plenty of spare high-pressure air available in the engine compressors which can be bled off into the cabin heating and pressurising system. Part-way along the fuselage, above the rear spar of each wing, there are sections of wall with no windows. Here air-conditioning ducts carry the air up inside the walls of the cabin to distribution points just below the overhead storage bins, to service the needs of the humans on board. The air exhausts out at floor-level grilles.

The cabin ceiling is a lightweight shell, non-load bearing, made of honeycomb-paper panels. Nothing much inhabits the attic space between ceiling and fuselage except large air-conditioning pipes and several supplementary vent fans. It is warm, stuffy and noisy.

At the very back of the attic space, in front of the rear pressure dome bulkhead with its silvery insulation lining like deep-buttoned upholstery, is the aft electronics centre. The angled sweep of the tail fin begins above. This back section of the aircraft is considered the most durable, and here, on a large tray, sit two bright orange-red flight memory boxes: the flight recorder, known as the 'black box', and the voice recorder. The flight recorder is not large — about the size of two toasters. A small, silver underwater location beacon is attached to its front. The voice recorder tapes all talk on the flight deck, but retains only the previous 30 minutes, so that as it records it also deletes. Two pieces of equipment for the Auxiliary Power Unit are also kept here: the battery charger and the electronic turbine control unit.

Behind the cream-coloured cabin wall panels, lagging next to the fuselage skin keeps the cold out and the warmth in. Each of the cabin portholes has three windowpanes. The inner layer is a scratch panel, but the middle window of acrylic could hold the pressure if the outer pane should fracture. Ultraviolet radiation has caused slight crazing on this outside layer. A tiny vent or breather hole in the middle pane prevents any condensation.

Where the plug of each cabin door fits against the fuselage opening there is less insulation. During flight a thin, cold draught seems to be coming in around the door edge. In fact, the pressure differential between the air outside and the cabin within is such that the door is held tightly shut while the aircraft is in the air. It is impossible to turn the large lever handle any distance towards the clearly pointing arrow to 'open'. The first opening movement of the door is inwards, and the pressure differential instantly works against this movement. The Flight Engineer makes certain the pressure differential is maintained at all altitudes. In an emergency, once on the ground, the door can be opened manually to a certain point and then a pressure pack takes over, pushing the heavy door fully around, flush against the fuselage side.

Down in the lower lobe galley two stewards heat up the hundreds of Economy Class foil-covered dishes of chicken and cashews prepared in the kitchens of Singapore Airlines. The modules containing the chilled food were loaded into the galley on the starboard side. Now the dishes are being moved at speed to ovens on the port side. First Class passengers are being tempted with a special Champagne Supper: caviar from the Caspian, Malay satays, Cantonese beef, fresh tropical fruit, coffee with Grand Marnier and cream, pralines and orchids. Business Class have the satays without the champagne and caviar. Passengers who have come on board at Singapore might be in need of food. Passengers from Australia are eating at 2.00 am Sydney time.

The lower lobe galley is an oasis of human activity tucked between the wheel wells and number 4 cargo hold. Hot meals are sent up in a small lift for distribution. Strict safety precautions protect any crew member in this lowest occupied level of the aircraft. There are three emergency exits, oxygen masks and fire extinguishers. The galley is monitored from the flight deck by the

Engineer and can be filled with inert gas instantly. It is a storeroom as well as a kitchen. Shavers are kept here in case passengers request their use, as well as spare alcohol — the 'back-up bar' — duty-free goods, bags of ice in refrigerators, long-life milk, take-off drinks and towels.

One and a half hours after leaving Singapore, food and drinks are cleared away and the lights are out. Passengers whose tickets have bought them space and comfort can lie with backrests lowered and legrests raised. They have bought width for their bottoms and distance from their neighbours. Passengers in Economy Class seats become acutely aware of their lack of personal territory. They doze, and sometimes sleep, bodies wrapped in blankets, a pillow placed wherever it might bring comfort. Some sleepless passengers want to chat; some babies need to cry. But the lights will stay off for five hours.

A steward in the lower lobe galley heats chilled meals for Economy Class passengers.

Outside, in the dark, the 600-watt outboard landing lights on the wings shine powerfully. 'See and be seen.' There are other aircraft around. The lights go on when the aircraft changes altitude; over the disturbed Middle East; when a Qantas aircraft is approaching from the opposite direction and the pilots flash at each other like truckies.

EBQ is lit in the same way as a ship, with a green light on the starboard wing tip, a red light on the port tip, and a white light at the rear, on the tail. The wing illumination lights are on. Red anti-collision beacons revolve on the belly and the top of the fuselage hump, flashing continuously from the moment the engines start, until they stop. Powerful strobe lights shining from the wing tips are able to pierce a greater distance ahead than any other light on the aircraft.

Inside the cabin the night is blanked off. Blinds are pulled down over each window like eyelids — 196 closed eyelids in a semi-sleeping plane. At the very back a very large man overfills a single seat, his body uncomfortably positioned on all available space. He is an 'oversize' passenger. All the early 747s carried a maximum of nine passengers across the body of the plane. The tenth seat was fitted in most 747s several years ago by making the armrests of each seat narrower, and by reducing the width of the aisles. The space between headrests is called the 'pitch'. An agreement in 1957 specified a maximum distance for Economy Class seats of 86 centimetres (34 inches), and 107 centimetres (42 inches) for First Class. Now airline companies make their own decisions, but most keep close to the old agreement. Business Class is somewhere in the middle. Humans come in odd shapes and sizes but the airlines do their best to make them all fit a standard space.

The 16-member cabin crew have now been together as a group since leaving Sydney on Thursday afternoon's QF1. Rosters are made up for each member of the cabin staff in 56-day slots, like the technical crew, but calculated on a different basis; the trip to London comes up, on average, once every three to four months.

The airline business has generated millions of jobs and created hundreds of new professions. Most are not visible to the general public. But the job of flight attendant is most visible. The eyes, the smiles, the moustaches and muscles, the figures and make-up of cabin crew are the focus of attention during hours in the air for hundreds of passengers. The technical crew are rarely seen, the four officers practically anonymous figures in naval-type uniforms with golden symbols and braid. The cabin crew are public.

The world's first airline stewards were three 14-year-old boys dressed like hotel bellhops. They were as small as possible to save weight, but Daimler Airways, who employed them in 1922, thought their presence on board would help sell tickets to passengers accustomed to servants. Later in the 1920s airline companies employed adult males, although a steward meant one less passenger could be carried. The job was similar to that of a steward at sea: 'to look after the well-being of passengers who might be taken ill', to provide rugs and maps and offer simple refreshments.

The world's first airline stewardesses had to be unmarried nurses, young and short. They had to dust the inside of the planes, bolt down loose seats, look after airsick passengers, serve meals and hot drinks, point out interesting sights through the window, carry a railway timetable in case the aeroplane was forced down in a field, and sit at the back on a mailbag or suitcase if all the seats were taken. They helped block passengers' ears with cottonwool to muffle the din of the engine, and unblock them by offering chewing gum during landing. They made sure passengers opened the toilet door, not the emergency exit.

The first stewardesses were employed by Boeing Air Transport in May 1930 on long inland USA routes. Many of the pilots — tough types who wore guns to protect the mail — did not want the women on board. But the stewardess in her white nurse's uniform helped passengers feel braver about bucketing along in an unpressurised, unheated biplane for hours and hours, 2000 feet above the ground.

The first European stewardesses were appointed by Swissair in 1934 and by KLM in 1935. But all aircraft flying long-haul routes carried male cabin staff only, because no airline company wanted the responsibility of the new-fangled nurse waitress on overnight stops. Ships carried stewards not stewardesses, and so, it was believed, should aircraft.

Qantas employed its first stewards on board the big Short S23 Empire flying boats which began service between Australia and England in 1938. Their first nine stewardesses were selected from 2000 applicants and began flying in May 1948 on the new Lockheed Constellations. They were 22 to 27 years old, not tall — 1.6 metres to 1.7 metres (five feet three inches to five feet six inches) — maximum weight 60 kilograms (nine and a half stone), and all were trained nurses or held a first aid certificate.

Age, height and weight requirements for flight attendants are not as tight today, but there are still nearly as many applicants for each job. The successful have to be robust because the work is physically hard, and preferably should speak a second language.

Flight attendants are servicing customers in a confined space for hours at a time. They must also keep a reserve tank of knowledge and training ready to react in a range of emergencies, many of which may never happen, one of which might have happened yesterday. Passengers must be made to buckle up, straighten up and stow possessions before every take-off and landing, while a relaxed, normal atmosphere is maintained. If an aircraft travelling at 120 knots along the runway should reject the take-off, the apple in a child's hand, for example, would become a missile, continuing to travel forward at the same speed. Emergency-procedures testing is part of every crew member's life and does not get easier with experience. The reactions have become instinctive. The procedures are so well known. But the test date comes up and the procedures have to be gone through again, and again. All crew members work within a certain level of stress because it is part of their training. Yet the driver of a family car, setting out on holiday, and almost by definition untrained in emergency procedures, is statistically much more likely to need them.

EBQ flies over the ocean towards India. Singapore Airlines is still ahead, by six minutes, flying at 31,000 feet. Swissair is still at 35,000 feet. So EBQ is still stuck at 28,000 feet, and is 35 to 40 miles to the left of track to avoid storms. Flight corridors over the sea are much wider than over land, but the pilot had to request permission from air traffic control to move so far.

This flight up through the bottom quarter of the Northern Hemisphere is considered the most boring of the three sectors. It is always dark. The aircraft accumulates hours of night time because it is moving against the clock. It will be only a little after midnight local Bahrain time when EBQ lands, although the flight lasts seven hours. The airport at Bahrain functions around the clock. International aircraft which have to fly schedules within other airports' curfews can land here when they wish. Not many passengers get on or off at Bahrain so the unsociable landing times are possible.

There are surprisingly few options available for arrival and departure times on the route between England and Australia. The late evening departure from Heathrow might seem sadistic to passengers weary from a day's activity. In fact, it is one of the only departure times which slips in between Heathrow and Sydney jet-noise curfews, yet gives a reasonable hour for passengers to board and leave at Singapore, a popular change-over point. Only seven hours out of the 24 avoid the pincers of the London and Sydney curfews on the QF2 route and practically all give unacceptable combinations of too early departure from London, too early arrival in Singapore or too late in Melbourne. Eight hours out of the 24

avoid curfews on the QF1 Australia–England journey. The first hour after the Heathrow night curfew is over, 6.00 am to 7.00 am, gives an arrival and departure time in Singapore which is reasonable for passengers.

Scheduling decisions, like catering decisions, are made in advance: November for the following April–November, and June for November–March. Scheduling decisions, like catering decisions, are always a compromise. It is possible to argue that the passenger's stomach drives all airline decisions. If passengers judge an airline company by the quality and frequency of food and drink, they book seats, which creates demand, which creates schedules. Or it could be argued that all flights begin with the passenger's pocket. If the price for a ticket is right, seats are sold and aircraft can be operated. But scheduling is a complex balancing act of organisation and prediction. Get the schedules right and an airline company knows what it is doing. The number of flights an airline decides to operate, the ports it decides to serve and the times of arrival and departure affect the number and type of aircraft it owns, the staff it employs, the amount of maintenance space it needs and its usage rate. One aircraft added to the Qantas fleet means, for example, 50 to 60 more cabin crew members to staff it and hundreds more hours of maintenance and overhaul time required in the workshops and hangars. A new route involves governments because governments exchange reciprocal traffic rights. British Airways operating long-haul out of Adelaide in 1982 meant Qantas, as part of complex negotiations and deals, was allowed to operate to Manchester. But the actual time QF1, for example, arrives at Heathrow has nothing to do with governments, or even the authority running the airport. It is horse-traded with the other airline companies. Only a certain number of aircraft can use an airport at once. A total of 120 airlines bring their schedules to a twice-yearly timetable co-ordinating meeting for 'trade a slot' time, with bargaining and dealing like any other commodity exchange.

The Fleet Utilisation and Scheduling Division of Qantas allocates operating patterns to aircraft plus their maintenance and overhaul timings. Qantas schedules work to a 21-day cycle. The plan has to be managed so that enough aircraft are available whenever and wherever they are needed, yet aircraft do not stand idle longer than necessary. An aeroplane is a long-term commitment of resources and it needs to be utilised as tightly as possible. Spare capacity has to be built in but not too much of it. A cancelled or badly delayed flight affects the whole carefully designed structure. Outside circumstances, like strikes or international tension along a route, cause great scheduling stress.

Fire fighters are trained to cope with airport emergencies.

In emergency procedures training, cabin crew are taught precise methods for dealing with emergency situations that could occur during a flight.

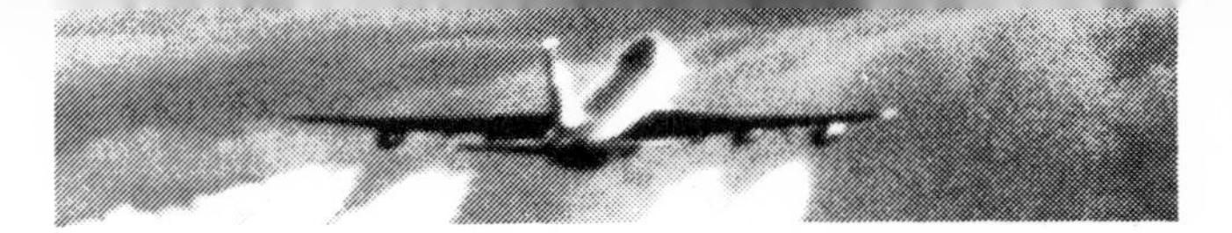

Aircraft run punctually to schedule because the customer is buying punctuality, but also because punctuality is essential to the proper functioning of the schedule. Standards of hygiene in aircraft catering are high not only because passengers require it, but also because an aircraft turning back in the middle of a long-haul flight with passengers suffering from food poisoning causes scheduling havoc. Engines and systems function as efficiently as possible for the same reasons: if there is too long a delay the knock-on effects are large. The sequence of meals falters. At Heathrow Airport, London, a restaurant waits, tables set, chairs empty, in case a planeload of passengers has to be fed an unscheduled meal. With a serious delay, crew members go beyond agreed working hours and have to be allowed statutory rests and sleeps. Accommodation for passengers may have to be found or passengers may have to be transferred to other airlines. A replacement aircraft may need to be summoned. The flight's profits shrivel as expenses accumulate.

The atmosphere on the flight deck is relaxed. The First Officer is resting in a little alcove behind the cockpit. The Captain has worked out a duty roster giving each member of the four-man crew two rest periods of 45 minutes during this seven-hour flight. Many airlines fly with only three technical crew on their 747s.

The Captain is sitting in his seat on the left, the Second Officer on the right and the Flight Engineer is in front of his panel of dials and switches. Both control columns move a little, under the direction of the autopilot. The autopilot cannot be seen. Only a lit-up switch indicates that this silent controller of the aircraft is at work. That and the fact that the Second Officer has his feet up and the Captain is making calculations on a piece of paper. The autopilot generally takes over about 10 minutes after take-off until a few minutes before landing — unless the landing is on automatic as well. Dial a new altitude on the autopilot control panel, mounted just below the windscreen and above the instrument panel, and the aircraft climbs to the level required. Select a switch and it adjusts to turbulent conditions. The autopilot uses a computer to process information from the INS gyroscopes, sensors and other instruments, about the course of the aircraft's flight. The computer passes on commands to compensate or correct the flight path by adjustments to the control surfaces on wings and tail. The 747 is a long aircraft and the autopilot is a sensitive monitor, sensing cues and acting before an error appears. It is really a stabilising system, performing the time-consuming job of controlling the aircraft's movements during flight. EBQ has two systems, although only one operates at a time, the other being

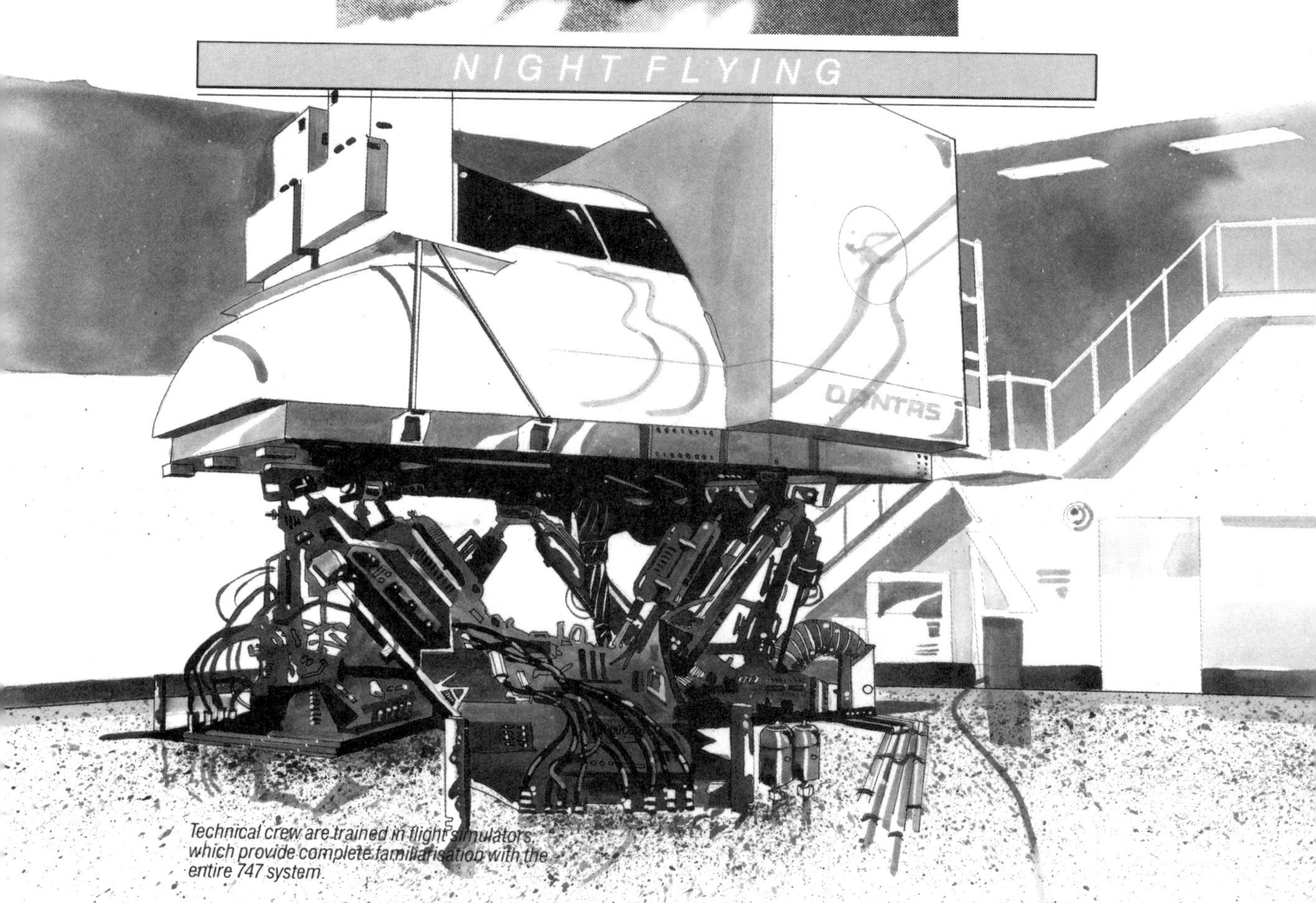

Technical crew are trained in flight simulators, which provide complete familiarisation with the entire 747 system.

ready as a backup. 'It flies better than we do,' comments the Captain.

Qantas flies the longest sectors of any airline company in the world. The return journey, Sydney–London–Sydney, all 37,000 kilometres (23,000 miles), involves, for example, exactly six landings, six take-offs. But three weeks ago the Captain landed at Bahrain five times. He landed in reduced visibility during a sandstorm, using ILS (Instrument Landing System). He landed with one engine out, in heavy rain at 6.00 am, and twice at night, hand flying. The landings happened over a 90-minute period. They were made courtesy of a simulator at Flight Training Centre, Jet Base. All pilots have to keep their licences current, and airport requirements vary: so many instrument approaches performed in the last 45 days, a night landing at least once every 90 days. Qantas pilots are permitted by the Department of Aviation to use the simulator to get their route qualifications and their 'recencies' to the required level. It is an efficient, cost-effective system. Simulating the real thing, plus a repertoire of emergencies, without using a thimbleful of that costly component of the real thing, jet fuel. Using an actual aircraft is highly expensive — 12 times more expensive than using the simulator. Jet Base's two newest simulators work 18 hours out of every 24, with the remaining six hours needed for servicing.

Simulators used to look like aircraft, from the outside. Inside, with a delayed response time, pilots pretended they were flying a real aeroplane. The latest simulators look nothing at all like aircraft, from the outside. But inside it is impossible to tell them apart from the real thing. Computer graphics surround the cockpit windscreens, giving the impression of being at any airport, in any conditions, landing, taking off, 35,000 feet over India at two o'clock in the morning. The pictures are created from thousands of points of light, rather like the dots in a newspaper photograph. Hydrostatic systems smooth the ride, create the skid of tyres on ice, the surge of take-off and the bumps in turbulent air. A 747 can fly with two engines out, even both engines disabled on the same side. The simulator summons up all the conditions likely to be met during a pilot's career, and adds catastrophes with unnatural ferocity. Controllers and instructors sit at consoles directing the action. The action is instant, the conditions are realistic and the risk — does not exist. No-one wants to blow sets of tyres in an actual 747 on a wet runway during high crosswinds just to test pilot reaction. In the simulator you can do this again and again and again. The latest simulators feel so realistic that pilots can leave a catastrophe session shaking and sweating. The simulators are so sensitive that a bad 'crash' can put them out for several weeks.

At a quarter past four on Saturday morning Sydney time, a quarter past two Singapore time, 1815 UTC on Friday, EBQ meets India at Madras, part way up the east coast. A steward brings coffee and orange juice to the technical crew. The only lights on the flight deck come from the instrument panels — muted greens and yellows, cream and orange and blues. There is a clarity to the night's darkness. The sky seems huge. During World War II, Qantas crews flew Catalina flying boats across the Indian Ocean, from Perth to Colombo, around 28 hours non-stop in the air. This was highly dangerous but vital flying, helping to create an air route to Europe.

It will take only one and a quarter hours to fly over India. Up here, in the middle of the night, there is no sense of India beneath. But even on the return journey, in daylight, the land below is not recognisably India. Monsoon clouds or a dust haze often hide its physical shape. The temperature outside is minus 23°C, 12°C warmer than usual. Fuel is being burnt at a higher rate because engines function less efficiently when the air temperature is warm.

EBQ left Singapore late. Now there is a feeling of being 'hemmed in' with Singapore Airlines in front and Swissair behind.

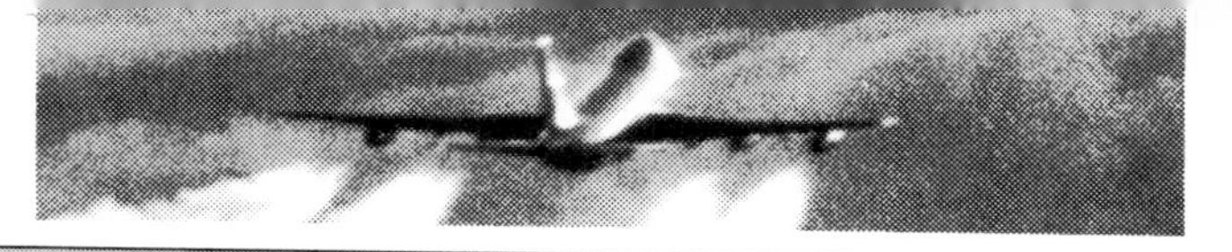

A decision is made to request permission from air traffic control to avoid Bombay and fly via the town of Belgaum further south. Air corridors are often organised in doglegs to pass over major reporting points. Avoiding the dogleg of Bombay will gain six minutes' flying time, cutting out 50 miles of flying. It will also give the opportunity to take advantage of Air Traffic Control Bombay's radar cover and request permission to climb to the level so long wanted, 35,000 feet, flight level 350. Bombay is a 'fix-it' area. Permission comes through. Dials on the autopilot control panel are set to the new altitude, and the mach number creeps up until the desired value is achieved and held by the autopilot. With the new altitude reached, vertical speed returns to neutral, and the thrust levers move back, without the touch of a human hand.

The crew talk about the old days, 10 years ago, when Qantas stopped in India. In four hours' time they will have 24 hours off in Bahrain. It will be hot, and there will be nothing much to do but rest and relax in the hotel. This crew left Sydney the day before EBQ on Thursday's QF1, and will arrive in London the day after, bringing in Sunday morning's QF1. The following Wednesday morning, around 5.00 am local time, they will be back in Bahrain for 24 hours, having brought QF2 (London–Australia) on the first leg to Bahrain.

They have two more long trips during this bid line: Sydney to Bombay and return, then over to Vancouver and return — 30 days' flying out of 56, with 157 hours 26 minutes' credit time.

Much of flying time is taken up by monitoring systems, a particular kind of mental discipline, checking and crosschecking, doing sums, looking at relationships, testing stated values against an innate sense of what they ought to be. Except during take-off and landing, the aeroplane is 'flying itself', and the 747 is a remarkably tolerant piece of engineering. But during cruise, a pilot, apparently relaxed, will be butting his knowledge and his training up against the continuously shifting reality, to keep his skills in trim. Pilots are always on standby. A pilot can be said to have been truly successful only at the moment of retirement. By that time he will have carried four million people, and delivered them safely to their destinations.

Flying at 35,000 feet, EBQ crosses into the Arabian Peninsula over Muscat, on the Tropic of Cancer. The tropic goes through Dacca and Hong Kong, through the middle of Mexico and the bottom of Egypt. Muscat is a small, rich oil city on the edge of the desert. Nothing can be seen of the desert down below in the blackness, but the flares from oil wells are visible, little golden blobs of light.

Like patients in a large hospital ward, passengers are woken

after too little sleep. The lights go up, the cabin staff present a 'refreshment' to fortify against the landing and waiting at Bahrain — tea or coffee, and biscuits. The only offering which is the same for all passengers.

On the flight deck the First Officer has decided to fly EBQ by hand from 'top of descent' to touchdown, from 35,000 feet to sea-level. The higher the aeroplane, the earlier the start. A tailwind, and descent begins further from Bahrain; a headwind, and it starts closer. The descent will take 30 minutes. The autopilot commonly does all but the final 1500 feet, but doing it manually is flying, and a flight deck decision.

A blue-covered *Aircraft Performance Manual*, the 747 bible, prints tables and graphs of the aircraft's performance under all conditions. Given this weight, height, air temperature and wind speed — that will follow. The First Officer works out the heights EBQ should be at set distances from Bahrain, and writes them down on a notepad kept below the window, by his right elbow. He commits himself to start the descent at 112 miles out from Bahrain.

At half an hour past midnight local time, 2130 UTC, the First Officer gently noses EBQ over. At this speed only the lightest pressure is needed to push the aeroplane's nose down the one or two degrees necessary. The autopilot disconnect button has been pressed, causing an alert to sound. Now the thrust levers are drawn smartly back to idle.

EBQ is dropping three-quarters of a mile a minute, far faster than an express elevator. It is very quiet in the cockpit, and dark, almost subdued. The First Officer concentrates on the instruments, balancing them against the descent profile he has worked out, checking height against distance remaining and time left to touchdown. Two instruments, the INS readout, and the DME (Distance Measuring Equipment) on the central display panel, show distance remaining to Bahrain, the figures winding down so fast that the smallest values are a blur.

Passing through 20,000 feet, the Flight Engineer calls the four items of the 'Descent/Approach checklist'. The Captain is in radio contact with air traffic controllers. The Second Officer establishes contact with the Qantas ground-handling agents in Bahrain, requesting parking bay details and advising the expected time at the terminal. All monitor the descent — extra pairs of eyes when they are needed.

EBQ is moving downhill now at half a mile a minute, at a constant speed of 300 knots. The First Officer makes minor adjustments to the speed to keep within the height and times he

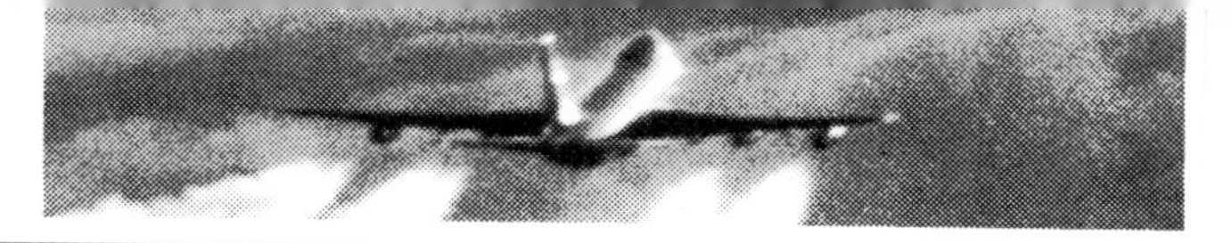

has chosen, raising the nose a little to slow the aircraft down, lowering it to quicken the descent. The speed for landing is calculated, given EBQ's actual weight. But critical speeds can be broadly spaced in a 747.

Thirty-five miles from Bahrain, with EBQ passing through 10,000 feet and jolting, the outboard landing lights are switched on, shining ahead like searchlights and making the aircraft more conspicuous. EBQ's nose is raised a fraction to begin reducing speed. Seventeen miles to go, and approaching 5000 feet, the pilot stabs at the cabin no-smoking and seat-belt controls and calls for the 'Before Landing checklist' from the Flight Engineer. The speed settles to 250 knots. At 3000 feet, 10 miles to go, the rate of descent slows to 700 feet a minute, and the speed reduces again, towards the approach speed, as the power is eased up to a landing setting. The wings have sprouted all leading edge devices, making it safe to fly more slowly. The wings are being artificially thickened, changing from an aerodynamically smooth shape suitable for fast high-altitude flying, to a hump appropriate to slow speeds at low altitudes. The flaps extend in sequence, the pilot's fingers feeling the physical shape of the flap selector lever on the central pedestal, differentiated from the feel of the other controls, so touch reinforces action. For each set of flaps thrown out the speed can be safely reduced by 20 knots.

At 1500 feet the pilot feels for the wheel shape of the undercarriage selector lever on the central instrument panel. Doors open, 18 wheels extend and lock into place, the doors close and a safe landing position is confirmed by three sets of green lights. The flaps roll to their final setting, creating positive drag now, and speed is reduced to a final 140 knots. EBQ approaches the runway at an angle of three degrees, sinking down. Landing clearance is through from the tower. At 50 feet, closing throttles. The control column is eased back to raise the nose a little and EBQ touches down. Then the nose angle is lowered so the nose wheels come into contact with the runway surface. Reverse thrust levers are moved up and back to activate the engines' reverse system, destroying their ability to thrust forward. Spoilers on the wings, acting as extra brakes to speed, are automatically activated to dump lift, helping the aircraft to settle positively on the ground so the wheel brakes can work efficiently. The pilot is still flying the aircraft along the runway surface. The speed reduces to 80 knots, then 60 knots. And EBQ is being steered along the ground, through the darkness of the night, and off the runway.

10. THE LAST LEG

The airport at Bahrain is busy, for one o'clock in the morning. Five big jets are lined up across the front of the terminal building: two Qantas (the other going on to Athens), British Airways and Cathay Pacific to London, Singapore Airlines to Frankfurt. Two of the aircraft have delays posted on the flight information board. The terminal seems filled with passengers. Some walk up and down stretching stiff legs. Some slump immediately onto seats. Some look at the shops, little windows into another world, or buy duty-free liquor. Away from the international arrival and departure gates, Arabs sleep stretched out on seats, waiting for flights which are not yet. Passengers are called to their aircraft and people detach themselves from seats, shops and walking, to cluster in their several hundreds at a departure gate. 'We belong to BA'. 'We are the ones going to Athens.' 'We are Cathay Pacific passengers.'

A new, fresh Qantas crew walk through departure control to EBQ. QF1's fourth and final crew. For haggard passengers who have been travelling non-stop from Australia it is now mid-morning on Saturday, Sydney time. The night has probably given them only a few hours' uncomfortable sleep.

The surface of the world is divided up into 24 time zones, each representing one hour, and 15 degrees longitude, although the lines dividing time zones do not curve up the world from south to north in neat segments but wiggle and twist, fitting around borders of countries, and including or excluding islands. The 24 hours must begin and end somewhere, and in 1884 the countries of the world agreed to accept a line passing through the courtyard of the Greenwich Royal Observatory by the River Thames, in London, as zero degrees longitude. Since the sun moves in a westerly direction, and time follows the sun, countries to the west of London are behind in time, and countries to the east are ahead. Flying across time zones, east to west or west to east, disturbs the body's natural rhythm, because body time and local time become separated. The unavoidable effects are known as 'jet lag'. The body needs time to adjust its own routine of eating, sleeping, and going to the lavatory, to a new local time, which can bring darkness, dinner and bedtime 10 hours earlier or later than usual. The first aviators to fly the Atlantic, Captain John Alcock and Lieut. Arthur Whitten Brown, noticed that they woke up very early feeling peculiar after their historic non-stop flight of 16½ hours from Newfoundland to Ireland in 1919. Probably in the future all passengers will have to put up with it, noted Brown. The effect lasted several days.

According to medical thinking, air travellers need at the very least one full day to recover from every five hours of time change, which means at least two full days for the Australia–England trip. Reactions are slower and thinking liable to fuzz after a flight of

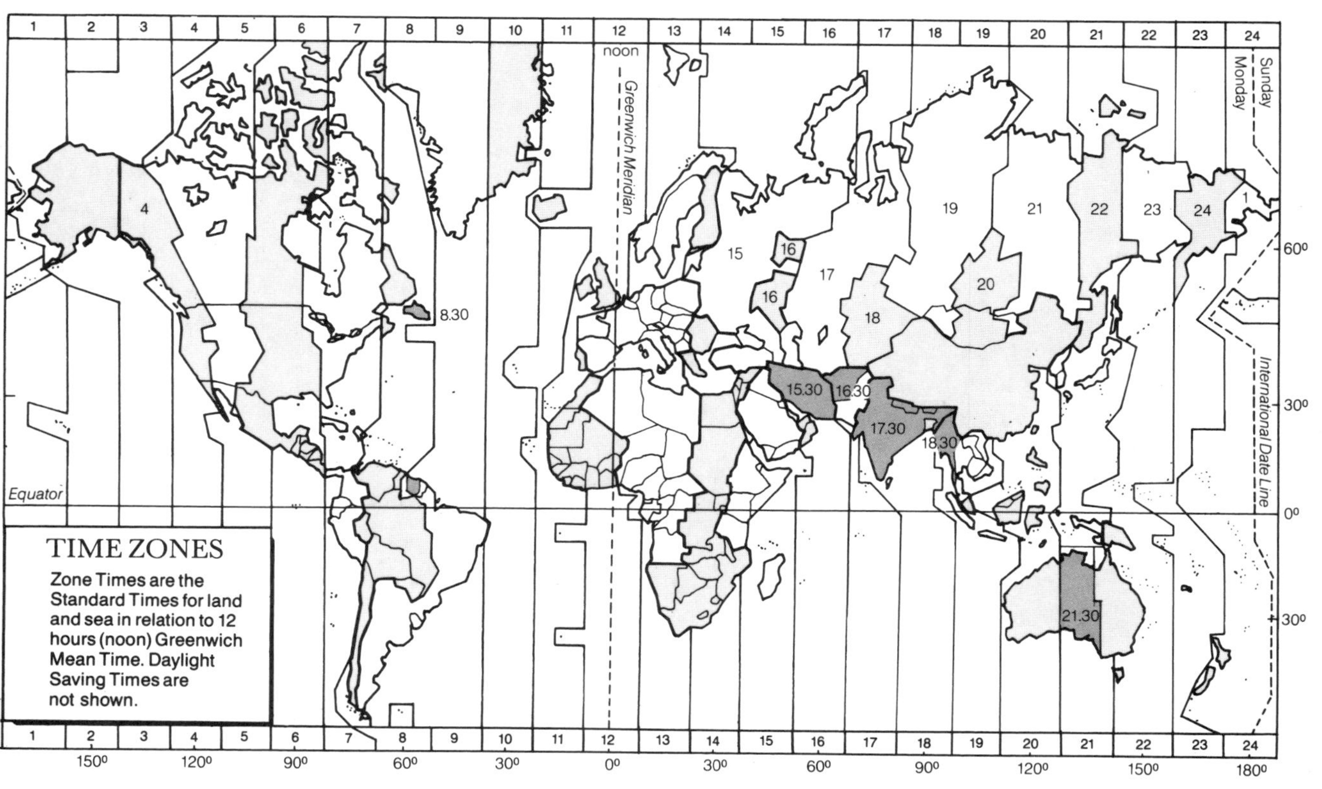

this longitude width. Certain things help to minimise the effects of jet lag: eating moderately, drinking plenty of non-fizzy and non-alcoholic drinks, exercising at transit stops, no smoking, breaking the journey if possible. Much of this is built into technical and cabin crew regulations. This crew boarding EBQ in Bahrain left Australia last Wednesday. They must drink no alcohol for eight hours before flying. They have the right to sleep for eight to 10 hours per stopover.

QF1 passengers are called to Gate 3, and return again to their seats. Seven hours to go before landing at Heathrow.

But one new passenger does not make it. The Captain, high up in the cockpit, watches a passenger making an attempt to get himself through the final control point. Considering the man unable to occupy his seat in the aircraft without inconveniencing other passengers, the Captain exercises his right, and refuses entry. Only seven people are joining at Bahrain and their baggage has been loaded into a small container, so it does not take many minutes to remove this man's possessions from the hold. A Captain can refuse to take off with unattended baggage, for security reasons. If a passenger does not appear but his luggage has been taken on board, a Captain, after considering all possibilities, can order the luggage to be unloaded. But to refuse entry to a passenger is very rare. This Captain has never done it before.

Passengers get their Qantas take-off towel to revive sagging faces and clean the airport from their fingers, and everyone is offered a drink: orange juice for Economy, champagne or beer for Business and First Class.

The minutes of delay build up. EBQ landed late at Bahrain. Congestion in the terminal building wasted some more minutes. A vegetarian meal has not been delivered but can be improvised by using eggs. The 'difficult' passenger wasted several more minutes. Turnaround time has been 70 minutes plus, instead of 60, and EBQ has to be off chocks by 2.12 am Bahrain time to make the slot time allocated by the controller. At 11 minutes past, the chocks are pulled back and the ground tug pushes EBQ tail-first out from the terminal building.

Taxiing seems to take a long time in the dark. At 2.25 am EBQ waits at the edge of runway 30. A Cathay Pacific plane lands, lights full on. Then at 2.30 EBQ moves out onto the runway.

'Qantas One into position and hold.'

Here in Bahrain, the runway is studded with lights, a city-full — red, green and amber lights embedded in the runway surface and encased in metal to withstand the thumping weight of 747s.

By now, in the middle of the night, they are very hot. If an aircraft inadvertently parks on them for too long the heat can damage the tyres. But sometimes in Bahrain it is necessary to hold for minutes on the runway threshold. Aircraft seem to fly around the world in herds. The Bahrain controller must allow 10 minutes' separation between aircraft in the airways above. He watches the traffic in the airways, waiting for an opportunity for EBQ to be absorbed, waiting for a space to push EBQ up into the system. Once in, at an altitude, a better altitude can follow.

The moment comes. Controllers must only use the word 'take-off' when they mean exactly that, otherwise misunderstandings could occur.

'Qantas One you are cleared for take-off.'

Out from Bahrain the cabin lights are dimmed. Some passengers want to go back to sleep. The Bahrain joiners are perky. One has just left an oil rig, another has finished contract work in Saudi Arabia and wants, above everything else, a stiff drink. Passengers will eat breakfast, see a film and have more refreshments to fill the six hours 50 minutes from Bahrain to London. The Flight Service Director decides to roll breakfast an hour after take-off.

The air in the cabin feels cool, and passengers hook rugs around their shoulders. Bodies loll, mouths sag open, heads droop. The senior flight attendant on the upper deck remembers the old days, 20 years ago, on Boeing 707s, when four stewards and one hostess looked after a maximum of 140 passengers. Five technical crew were needed, and the hours were long with no overtime. Food came on board packed in boxes. Stewards set the trays and handed them out two at a time, balancing the glass of iced water and the bowl of soup. The job, he says, is much easier and simpler today. Shorter hours. And no soup.

Abeam Kuwait, EBQ has reached 31,000 feet and is climbing. British Airways to London is ahead but too far to be of interest. Singapore Airlines to Frankfurt is 10 minutes ahead, and a Qantas plane to Athens is only just ahead but once past Damascus will be changing track. A 30-knot tailwind is diminishing EBQ's delay. But the aircraft is heavy: full in Economy Class, except for the seat left spare by the non-boarding Bahrain passenger; eight babies; 27 First Class; 35 Business Class. There are an extra three hours of fuel in the tanks: one and a half hours of contingency fuel, and another one and a half hours because a fuel strike is threatening in London and the Captain has authorised enough extra fuel to fly EBQ from London to Amsterdam. It costs fuel to carry fuel and all decisions to add extra are made carefully. In flight, each engine drinks 60 litres every minute.

Two air chefs dedicated to First Class stomachs cook their charges' breakfasts: eggs, scrambled or boiled, bacon, sausages, tomatoes, fish. The eggs, along with the yoghurt and the tropical fruit in a decorated basket, were put on board in Singapore. The other ingredients, and the orange and grapefruit cocktail in a champagne glass garnished with two mint leaves and half a cherry, came on in Bahrain. Economy Class have stewed fruit, a tomato omelette with bacon, a Danish pastry, and tea or coffee. Business Class have all this and fresh fruit, and a couple of cocktail sausages as well. Breakfast is a nice straightforward meal. People seem to need it. After all, nine hours have passed since supper out of Singapore.

It is movie time. The films have been shuffled around the aircraft so each screen shows the film it didn't show out of Sydney.

At 0200 UTC, EBQ has crossed the coast of Turkey, flying at 35,000 feet. The route lies east of the Mediterranean, over the land connections between the Aegean and the Black Sea, up across eastern Europe — Bulgaria, Yugoslavia — and into Germany.

At 0226 UTC on Saturday morning, dawn breaks. The sky here above the Sea of Marmara is very pale and light. Darkness since Bahrain has lasted only three hours. The European day is long at this time of year. But darkness for the through passengers began as they left Australia. The crew are happy. Flying uphill, towards London, the sun is not in their eyes. Passengers are locked into flickering movie screens and most are unaware that dawn has happened.

At 0241 UTC, EBQ is over the top of Istanbul and within Istanbul Air Traffic Control. Skies are crowded over the Middle East and Europe, and flight corridors are narrow. There are many flight control requirements: reporting in at the boundary of each country, conforming to military demands. Eighteen flight control stations order the route between Bahrain and London, keeping technical crew busy. They check waypoints, distances and direction, and make contact with air traffic controllers at the boundary of each zone. Controllers can work horizontally, informing a neighbouring controller of an aircraft's needs before that aircraft arrives over the new space. But sometimes controllers will not speak to each other. Political realities dominate international air, and control zones often parallel national boundaries. The quality of air traffic control varies. It is an exacting job needing well-maintained modern technology.

At Radovets, EBQ passes into Bulgarian airspace, the first of the Eastern Bloc countries, crossed in 20 minutes.

'Good morning Qantas One maintain level three five zero. No

further contact till the boundary,' says a voice in heavily accented English.

English is the international language of air traffic control, but to simplify communication between pilot and controller, of whatever nationality, a special shorthand vocabulary has been developed. It minimises misunderstandings and cuts down the number of words necessary. It has the attraction and mystery of any in-language to the eavesdropper. But pilots flying internationally must learn to tune their ears to the various pronunciations of 'aviation English', and to interpret the intent through the accents, from Japanese to Hungarian.

Around 6.00 am Bahrain time, 0300 UTC, the 8-mm film projectors stop whirring, and window blinds are pushed up. Below is Europe. Even the weariest passenger is stirred. Ahead is home, a holiday, a new job, an old life, or at the very least the end of the journey.

The Flight Service Director decides to run the bar around. Passengers contemplate the length of the toilet queues. There is a flurry of duty-free selling. People crowd around the senior flight attendant, with his boxes of perfume and cigars and cigarettes. The gold-plated koalas are going fast — he has sold 71 — and 19 of the silver-plated Australian flags, each with an opal tucked into its curve. Suddenly, to some of the passengers, this aeroplane seems like home. Ahead is strangeness. Different money, shopping worries. They push notes into the attendant's hand and he stuffs money into an overflowing box. He is down to one Chanel No. 5, no Fidji, very few cigarettes. He is responsible for the cash from sales during the whole 11-day trip, and must get it back to Sydney with a correct record of stock sold.

Refreshments are handed out. 'Morning tea' for Economy Class is a shortbread, a cup of tea or coffee, a bar of chocolate to slip into the handbag, and the last of the Australian fresh food on the aeroplane, an apple, banana or orange loaded yesterday in Sydney. First Class have a meal if they want it: soup, savoury, tropical fruit from Singapore and stuffed dates from Bahrain. Business Class have the Economy Class assortment but their fruit comes in a basket and they are offered a savoury flan.

Forty minutes from touchdown, all food and drinks, all paper napkins and orange skins should be cleared away. The Flight Service Director adds up the wheelchairs needed at Heathrow and gives the figures to the flight deck. Half an hour out from London QF1 can contact the Qantas operations room at Heathrow by radio link-up and the requirements will be sent through. The Flight Service Director has flights to Hong Kong, Honolulu, Los Angeles,

During any one sector of a Qantas 747B flight over 1000 meals may be served.

Auckland, Bombay and Jakarta in the next two months, but he wants a new sports coat and he plans to buy it in London today.

Duty-free selling will stop at least 30 minutes from London because the senior flight attendant in charge must stocktake and make accurate declarations for customs at Heathrow. While other crew are applying their aftershave lotion he will still be checking totals.

Passengers tidy their belongings and continue to queue for the toilets. Economy Class is like an over-large family getting up in the morning.

Up on the flight deck it is bright, and the sky is a light pale blue. EBQ begins a gradual descent over Germany. Heathrow is 340 miles and 55 minutes away. Europe looks soft in the early morning. The crew are happy. They left Sydney on Wednesday, and they have 63 hours 20 minutes of rest ahead of them in England. The First Officer will visit his parents; the Captain will hire a car and drive north to see friends.

EBQ leaves the continent of Europe at Ostend. It is 06.19 hours local time, 0519 UTC, and the air temperature outside is minus 53°C. Three minutes later and over the Channel, the sea is a wide, wide ribbon dividing two pieces of land. Out the Captain's side window, on the left, Calais is visible, and the white cliffs of

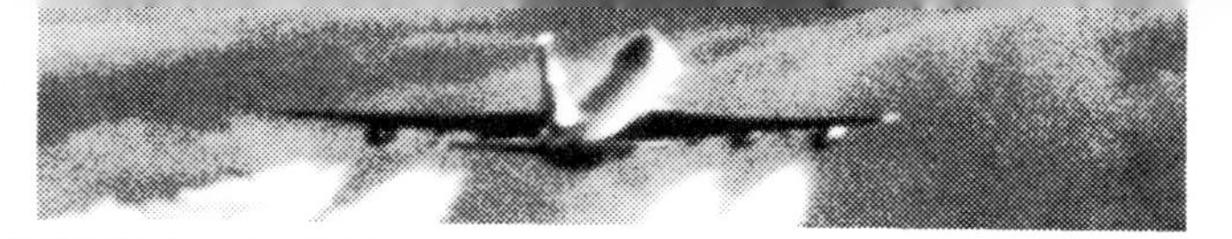

Dover, with splinters of ships on the sea between. English air traffic control based at West Drayton, 10 kilometres (six miles) north of Heathrow, takes over when EBQ is mid-channel. There is a small ripple of relaxation. EBQ has almost reached the far end. English air traffic controllers symbolise many things. And there is always time for a cricket score, if there is one worth giving.

EBQ turns a little, towards Clacton on the Essex coast. Straight ahead the wide estuary of the Thames is like an entrance. EBQ flies on, up the line of the river to the thickness of London. The First Officer is hand flying the plane now. EBQ is low — at 6000 feet. Twenty miles and nine minutes from touchdown. Down to the right the Thames shines among the buildings. Tower Bridge shows clearly — then it's gone. The Captain has taken over. Flaps out, and descend to 2500 feet. Then — in the distance but directly ahead, looking like a giant, grey, chopped-short road — is the runway. EBQ comes in, it seems, on the same straight line since the Channel. At 06.43 hours landing gear down. Two minutes later and touchdown, smooth and easy. The 747 slows.

'Gate M 28,' notes the Captain, and turns off the runway to the right.

'Good morning,' says ground control. 'Straight ahead please.'

EBQ trundles along with a gentle bumping motion. Someone checks the Heathrow Airport map. The Captain steers by turning a tiller near his left knee. He drives past the tails of parked planes, turns right again and very slowly and gently takes 260 tonnes of aircraft nose-first up towards a wall. Two green vertical strip lights, set in the terminal building at cockpit level, indicate straight, accurate docking. If the 747 strays towards the left the left strip shows red; if it steers to the right the light on that side turns red. A separate display board, like a scoreboard, has a white-painted line for each make of aircraft. As the line indicating a 747 centres above a white neon light on the display board, the Captain stops.

QF1 has arrived. At 06.53 hours local time, on a Saturday morning in London. Two minutes ahead of schedule. VH-EBQ *City of Bunbury* has now landed a total of 2094 times, and flown 22 hours three minutes since leaving Sydney, giving it a total of 10,147 hours 33 minutes in the air.

11. TURNAROUND

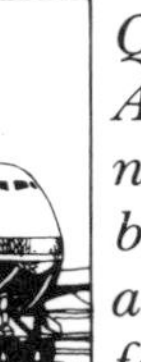

Qantas Flight One is the fifth flight Heathrow Approach Control have brought in this morning. The noise curfew was lifted 55 minutes ago. The busy time begins around 6.30 am and continues until 11.00 am, as aircraft arrive from all over the world, and depart for Europe. Qantas likes its 6.55 am arrival slot. Only 'grandfather rights' (being around for a long time) ensure a good time like this.

As soon as EBQ touched the runway, the compression of the tyres triggered off a separate electrical circuit which only operates while the aircraft is on the ground. Now it is possible to open the cargo hold doors and use the rollers set in the hold floors to move containers. Now the voltage in the three drainage pipes in EBQ's belly changes because it is not necessary to heat the pipes as much — the air can never be as cold as at high altitudes.

EBQ docks, the chocks go in under the wheels, the airbridges reach out and the cabin doors are opened, from the outside. All doors have been put to manual. While the aircraft moves they are on automatic, primed like a cocked gun, ready to fling out their safety equipment at enormous speed and strength. Emergency chutes and slides are stored in the thickness of the doors, folded into one hundredth of their potential size, with pressure packs ready to help inflate them in a maximum of eight seconds. Now the doors have been 'disarmed' — they are passive. The wing slides packed inside the fibreglass wing fairings will not deploy.

Air Canada are the Qantas agents for handling passengers, cargo and baggage in London. They own the specialised heavy equipment which is lined up ready to move in and start unloading.

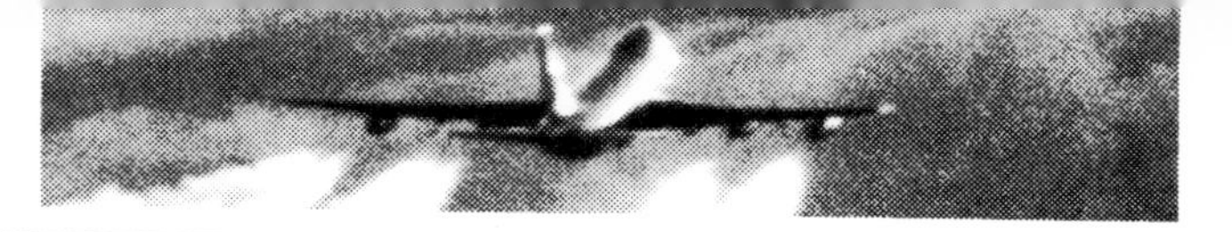

Within minutes of chocks going on, a cargo handler is pressing buttons on the control panel of hold 4 and rolling the priority baggage container out. A door swings open on main deck right and the catering truck looms outside, a platform extended to the aircraft ready to take out the galley carts. A Qantas duty officer checks the time the first baggage container leaves the aircraft: 7.03 am. He has already radioed the Qantas Flight Operations room in the Terminal Three arrivals building, reporting the on-chocks time so Sydney Operations Control can be signalled that the flight is over.

The first passengers begin disembarking just before 7.00 am local time, gathering their wits and their documents, to face the queues at immigration and passport control; to wait for their suitcases to appear, and haul them heavy off the moving belt, and go through customs, and emerge, to the real world of people who have not just flown across the world since yesterday afternoon, except that it isn't because it should now be tonight and instead it is breakfast time. Again.

EBQ is being emptied fast. The honey cart drains each toilet's holding tanks through pipes leading straight down into the large tank on the back of the truck. The load will be emptied at Heathrow's sewage farm, a truly international destination. A highly concentrated blue-coloured disinfectant is pumped with some water back into the holding tanks. Intercontinental germs cannot survive this treatment.

All baggage is gone from the holds by 7.21 am and all cargo is at the warehouse in the cargo terminal within an hour of landing. A customs officer checks through the cargo manifest. Less than five per cent is retained for inspection, and most customers can collect by midday.

One customs officer does board EBQ, to check with the senior steward that the bar and duty-free goods carts are locked. The quantities left must be correctly entered on forms, and the declaration will be dropped off at customs. The sealed bars and duty-free goods carts are taken to the bonded warehouse at the London caterers, where they are checked by customs staff. Any incorrect declaration and the goods are detained. Tonight, when EBQ leaves Heathrow for Sydney as QF2, First and Business Class passengers will drink duty-paid alcohol after boarding. The duty-free bar cannot be opened until the flight is underway.

The water tanker arrives with a top-up of London water for EBQ's three potable water tanks. Once a month the local health authority checks the quality of water in the Air Canada bowser. The much-filtered London water is added to the mix of Bahrain, Singapore, Sydney and Melbourne water already in the tanks —

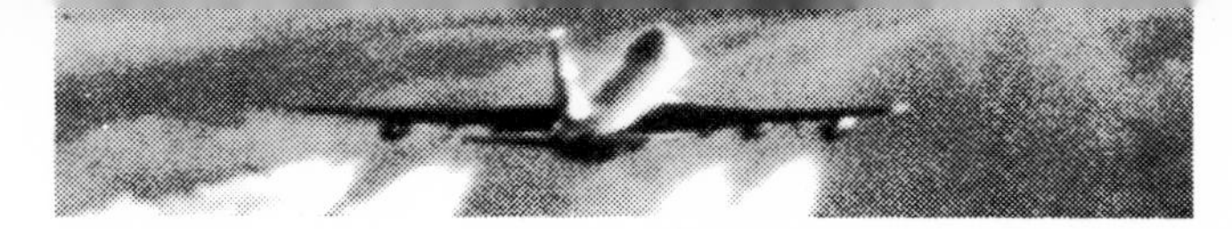

well travelled, expensively carried H_2O, kept sweet by added chlorine. The tanks get drained in Sydney and new chemicals are added.

British Airways engineers under contract to Qantas carry out routine maintenance plus some more, because this stop lasts over two hours. Engine oils must be checked within half an hour of arrival, while they are still warm and circulated. Here in London, magnetic plugs inserted into the engine oil are inspected for signs of any particles of metal in the system. The plugs are wiped on filter papers, any problems are dealt with, and the papers travel in a folder back to maintenance in Sydney, as a monitor of the engines' health. Tyre pressures must be checked in London, at least six hours after landing, and topped up with nitrogen if needed.

In the baggage claims area a tired elderly passenger lifts a suitcase which looks like his off the moving belt, but later on, away from the airport, he finds it isn't. Last Saturday's QF1 arrived in London with a lady who could speak no English, still in her seat when she should have got off in Bahrain. She had not recognised her stop and had to be returned on the first available flight. Some passengers have connections to make. Others wonder where all the Qantas people are, and worry that they might have made a mistake and be in the wrong place. Air Canada staff are available to help, and a Qantas customer service officer is on duty, checking that First and Business Class have priority baggage handling.

Around 8.00 am the last passengers have gone. The duty officer goes back to his desk to write an Aircraft Arrival Record, noting times and events in case of queries or complaints. He sends telexes if there are any problems. He has noted the fuel usage rate of EBQ's four engines in the cockpit Deck Log and the figures are sent back to Sydney by Flight Operations.

Technical crew and cabin crew go through a crew customs control and quickly disperse to their English weekends. If customs staff wish to examine passengers' baggage, it is done behind the scenes with sniffer dogs checking suitcases as they come out of the containers.

At 8.00 am a tug tows EBQ away from the stand at the terminal building. The space is needed for another aircraft.

All day EBQ parks out in the '150s', beyond the taxiway, standing amongst the world's aeroplanes at the edge of the airport. Parking fees for the day are high, the equivalent of three excursion Australia–England return fares. Qantas must also pay the British Airports Authority a landing fee, a navigation fee, a departure fee for each passenger to cover security checks, and a fee for the airbridges. An agent is paid for moving baggage between connecting flights and a Qantas flight. Air Canada and British Airways receive their fees for acting as Qantas London agents, and an American company, Marriott, is paid for dealing with food and amenities.

All catering for Qantas flights leaving London is done about 15 minutes' drive from the airport, in a building which used to be a nunnery. Where nuns once ate their frugal dinners, medallions of lobster and poached turbot, saddle of lamb and raspberry flan are prepared for 17 First Class, 26 Business Class and 207 Economy Class dinners for QF2 passengers leaving London for Bahrain at 9.15 local time on Saturday evening. Breakfast for Economy Class passengers to eat out of Bahrain is also prepared here.

Rural Hounslow is still just visible. Old hedges trace the shape of fields. There are ducks next to a picturesque black-timbered pub. But the support services for Heathrow Airport spread wider and wider around the original terminal buildings and runways. Like all major airports, Heathrow is a mighty employer. At least eight caterers vie to supply food for the 300 flights, on average, which leave Heathrow each day. Airlines spend $A100 million a year on catering at Heathrow. At around 5.00 am each morning the raw materials start being delivered here where airline service buildings loom near pre-airport Hounslow.

Charter companies tend to provide passengers with frozen meals, but scheduled airlines use cook/chill techniques to feed their passengers. So catering is a matter of daily logistics: how many of each type of meal to cook for whom by when. Meals are cooked in advance, as in most earthbound restaurants.

Airlines have their own specifications. Qantas, for example, uses no tinned vegetables. It takes eight seconds to make each Qantas Economy Class breakfast omelette, longer than another long-haul airline whose meals are being prepared at the same time, because it must be stuffed with tomato. Hot cooked food is put into a room-size refrigerator and cold food into another, and all the food is chilled down to 4°C for about six hours. The various items are then brought out and arranged on trays with the appropriate quota of cutlery, napkins and packeted biscuits, via an assembly belt, and loaded into the correct galley cart to be chilled again in another refrigerator room, the 'flight holding fridge'. Unlike the huge Flight Catering complex in Sydney, this kitchen operation is not computerised.

Carts filled with the remains of old meals, off-loaded from QF1 and other airlines' flights, arrive around the back of Marriott in the morning. Hotel-style washing up restores dishes, glasses and cutlery to the assembly lines for the new meals. Everything needed to re-equip the aeroplane is caged in a store section of the building and collected together every afternoon to go on board for the evening's QF2 flight. EBQ must leave London with the same number of Kleenex tissues, magazines and wooden toothpicks as it set out with from Sydney. The goods must flow backwards and forwards across the world in even quantities, like the tide. But at least the glossy magazines can go on board for passengers' idle moments, unedited. Qantas does not require the bra-and-panty advertisements to be ripped out, as do certain other airlines flying out of Heathrow.

Not far from EBQ's daytime parking space, next door to unkempt government buildings where Barnes Wallis carried out his research into the Dambusters' bouncing bomb, stands a modern windowless building. Inside, long racks of cabinets belonging to the central cargo processing computer system ACP80 keep track, for 24 hours

a day, seven days a week, of all air cargo as it is moved through Heathrow Airport, between the aircraft belly and bonded warehouses, through customs clearance, via fork-lift trucks to freight agent vans and lorries. Air cargo is a high-pressure business where time is the selling point. Air freight is used for high-value and time-sensitive commodities — computers themselves, video cassette recorders, boxes of ripening mangoes, blood serum, bull sperm. The business at Heathrow is very competitive, with 400 agents competing to get the orders from thousands of exporters and importers, and 74 airlines competing to get the business of 400 agents. Official IATA rates are published for air freight, but QF2 tonight will be carrying cargo beneath its 250 passengers at around half the current rate. Humans travel a great deal more expensively than cargo to Australia.

The catering and amenities for Qantas flights leaving London is handled by Marriott In-Flite Services operating from Hounslow.

Her Majesty's Customs, in peaked caps, sits in the middle of the air freight business like a giant spider, clearing consignments sight unseen, automatically via computer instructions, or demanding a consignment to be opened up by airline freight shed staff, who must then laboriously repack it to avoid damage claims. HM Customs sits astride the No Man's Land between airside and landside, controlling the crossing of goods and people, but much more interested in imports than exports.

Close to the roar of take-off on runway 10 left, a freight agent beyond the perimeter fence is labelling boxes of Rank Xerox equipment, wire-strapped to wooden pallets, for delivery the day after tomorrow in Melbourne. He has reserved cargo space on QF2 from his computer terminal. The consignment will be taken by van to the Air Canada freight sheds within the airport fence.

Airlines like receiving cargo already palletised or containerised. Agents who 'stuff and unstuff' save the airlines time and energy. Most of EBQ's allocation of containers are built up during the afternoon, ready to be taken to stand 28 for loading. The estimated weight of cargo ought to be sent in to the load controller by zero minus three hours, and the final weight by zero minus 90 minutes.

At zero minus nine and a quarter hours — midday — passengers can begin checking in for QF2. Some people have had to sign out of their hotels. Some have travelled long distances and are already at Heathrow.

Still in their hotel, tonight's crew are getting edgy. Bags are packed, shoes polished. Now they must try to sleep. The Flight Service Director has done his paperwork for the trip. Sixty-three hours seem a long time in London, but on Thursday after landing he slept all day and only began using his slip time after waking around 5.00 pm. Friday was a normal day, although he was still a little jet lagged. It is difficult not to keep waking at 3.00 am and the hotel now runs video films for Qantas crew in the early hours. This morning he did some last minute shopping. Now it is half past 12 on Saturday afternoon and he will sleep until the 6.00 pm telephone call, to be ready at 7.00 pm for transport to the airport.

At odd times during the day vehicles and people arrive at EBQ's side. Engineers deal with any snags noted by the Flight Engineer between Sydney and London. An Air Canada 'grooming team' of eight to 10 cleaners vacuum and wash down the interior, empty ashtrays and pick up rubbish, clean the galleys and find lost possessions. It takes around two hours to clear away the evidence of 400 people's occupation. Once every three months in Sydney the inside of a Qantas 747 gets a deep clean, three days' worth of foaming and scrubbing.

In the Qantas Flight Operations room computerised meteorological information is being printed out by the armful, swathes of paper crossing the room. Qantas buys weather forecasts from two international organisations, one British and one American, who have divided the world between themselves. 'Blockettes', squares with five degrees latitude by five degrees longitude, are on order to cover the England–Australia route, and a computer prints out the updated information. Other information comes in from meteorological stations and airports, both forecasts and actual weather: winds, speeds, temperatures, visibility, cloud base, any CAT (clear air turbulence) and CB (cumulo-nimbus) – for all altitudes.

QF2's flight plan is stored in a computer and is automatically activated every 24 hours into the London Air Traffic Control Centre at West Drayton. A controller sends it down the line to all air traffic control centres concerned.

Around 4.00 pm local time the file for tonight's QF2 starts to build up. Large numbers of 'Notams', Notices to Airmen, have accumulated in the Heathrow Air Information Service office, and the flight operations officer on duty goes through them, choosing the bits relevant to QF2's route. Maps of the weather come from the Civil Aviation Authority office. The Fuel Flight Plan cannot be constructed until a preliminary 'zero fuel weight' comes in from the Air Canada load control office. This figure, based on the aircraft's empty weight plus the estimated payload, should be available by 7.00 pm local time, but it can be late. The actual figure may not be known until 10 minutes before push-back, so the Fuel Flight Plan is based on an informed estimate. Should the zero fuel weight end up varying by more than 1000 kilograms from the planned figure, a new Fuel Flight Plan would have to be made.

Around 5.00 pm EBQ is towed back to gate 'Mike 28'. QF1 comes into gate 28 and QF2 leaves from 28 almost as a matter of course. It is near Air Canada's Maple Leaf Lounge for First and Business Class passengers, and by Air Canada's load control office. QF1 is the first aircraft movement Air Canada deals with in the day and QF2 is the last. Stand 28 is large, and 747s need space for the vehicles which must service them. It is Qantas' usual position, and the controllers in their darkened room, sitting in front of closed-circuit TV screens which show taxiways and the state of the stands, allocate it to Qantas practically every time.

The long tube of aeroplane metal has heated up during the day. All entrances are open but the air inside hardly circulates.

The cargo holds are empty and rather grubby, with bits of rubbish caught below the floor systems. The lower lobe galley is revealed for what it actually is — cargo hold 3, a space waiting to be packed around with galley modules, creating a room. The cargo holds here, in the belly of the aircraft, are like the cellars of a house, but the ceilings are low. No cosmetic panels cover the structure of the fuselage, and the skin of the aircraft shows in the places where there are no pieces of insulating wadding. At the back, in cargo holds 4 and 5, the walls curve in steeply. Bulk hold 5, at the very back, is small, with a sharply sloping floor, and the walls are lined because this is where loose cargo is stowed. During flight the temperature of the holds stabilises at around 5°C, but 4 and 5 can be heated to 15°C and air circulated, if animals are to be carried. But if animals are on board, certain categories of cargo cannot be stored in the same hold — dead bodies, for example.

Inside the cabin the plastic coathangers wait for First Class and Business Class coats. Heavy, narrow, silver-plated coffee pots and sugar bowls have stayed on board EBQ, with the wicker wine baskets, and the headsets supplied in Sydney for the complete trip. Everything else will be loaded into the cabin with the food. Although this is a turnaround flight, in most ways it is like an originating flight.

Yellow lights flash intermittently on the control panels of the flight deck, and the sunlight comes in through the open emergency hatch in the roof. Zero minus four hours to departure.

At around 6.30 pm a Marriott truck brings the food to the aircraft, in dry ice now to keep it chilled, and another truck carries the amenities.

The cabin crew leave their hotel for Heathrow at 7.00 pm. The Flight Service Director stands at the front of the bus like a guide on a tour, delivering a briefing on the flight to come and a pep talk on being a crew again after the days away from work. On arrival at Heathrow he will get out at the Passenger Terminal to begin his walk-around, and the rest of the crew will go through security and immigration and out to the aeroplane. There is a lot to put away and fix up before the passengers come on board.

At 7.15 pm the Qantas Duty Officer collects his walkie-talkie and telephones the speaking clock to check his watch. Zero minus two hours and he is working to local time. Heathrow is part of the Qantas global office. All day telex machines clatter messages, and the computer prints out information and instructions seconds after they are generated in Sydney. The paperwork for the flight is nearly all done. The Customer Services Officer has left to circulate in the terminal, checking on the passengers. The Duty Officer leaves to be physically present wherever he might be needed —

airside with the baggage handlers, with the cargo, at the Air Canada load office, checking with the caterer that the meal numbers are correct, liaising with EBQ's Flight Service Director and the technical crew. He roves via his car in case of emergency and for two hours he watches that everything is running smoothly and to time. His job is the same as that of the Flight Despatcher in Sydney, the Ramp Co-ordinator in Singapore. He carries the responsibility for filling EBQ.

Air Canada staff are processing QF2 passengers, and their specialised airport equipment lifts containers into EBQ's hold, according to a plan initiated from computers in Sydney and refined here in London.

QF2 passengers check in at the Air Canada counter at Heathrow Airport.

Heavy containers rise from ground-level, move forward into cargo holds and roll into position. The catering modules are already in the lower lobe galley locked into the floor, and the first baggage containers are being loaded into cargo hold 1 and 2. Now the body of the catering truck is poised five and a half metres (18 feet) in the air, unloading the food for the First Class galley. A small lift in the main cabin galley takes carts upstairs. Near the back of the aircraft, a metal container stands empty on the ground. Crew baggage and company papers go into this 'service container'. The bottom section is a built-in metal trunk, the diplomatic locker. No red leather, and not many diplomatic papers these days. In the front half go small loose items like fruit knives, or large scissors revealed in a passenger's hand baggage by the security X-ray, or perhaps a child's water-pistol, or a fragile small parcel. The back half of the locker is for company papers, and any diplomatic papers and valuable cargo. The Duty Officer must personally put the items in here, must be witnessed and must sign that they are loaded. Only he has the key. At the other end the Duty Officer has another key and must sign for any objects removed.

Ground engineers are going through the departure servicing routine. EBQ needs electricity and it is supplied by a power unit at the side of the parking bay. Yellow cables cross the tarmac to a connection point behind the nose wheel.

The engineers check that EBQ is an aircraft which is available to fly. They check outside for any dents that might have been made during the day, look down the engine pods for any sign of damage and through the interior of the aeroplane for a quick nick-check. Some of the aircraft's systems have been dead all day because the power has been off. Now an engineer resets the circuit-breakers on the flight deck.

At the load control office EBQ's load sheet is being built by hand, not by computer. The load controller has estimates of the total weight of passengers, cargo, mail and courier mail, and baggage. But he must wait for the exact totals of everything.

At zero minus 90 minutes the Duty Officer goes to the Qantas information desk in the Departures hall to find out the exact number of courier bags for tonight's flight. Four of QF2's Economy Class passengers will be couriers, as is the case every night, their personal baggage consisting of anything from eight to 40 large, bright-coloured plastic duffle bags tied with cord. Inside will be company papers, contracts, computer tapes, publisher's manuscripts, smallish engine parts — whatever needs to travel fast. The bags are not individually weighed. Each is reckoned at 30 kilograms, and Qantas charges an agreed amount for all this excess

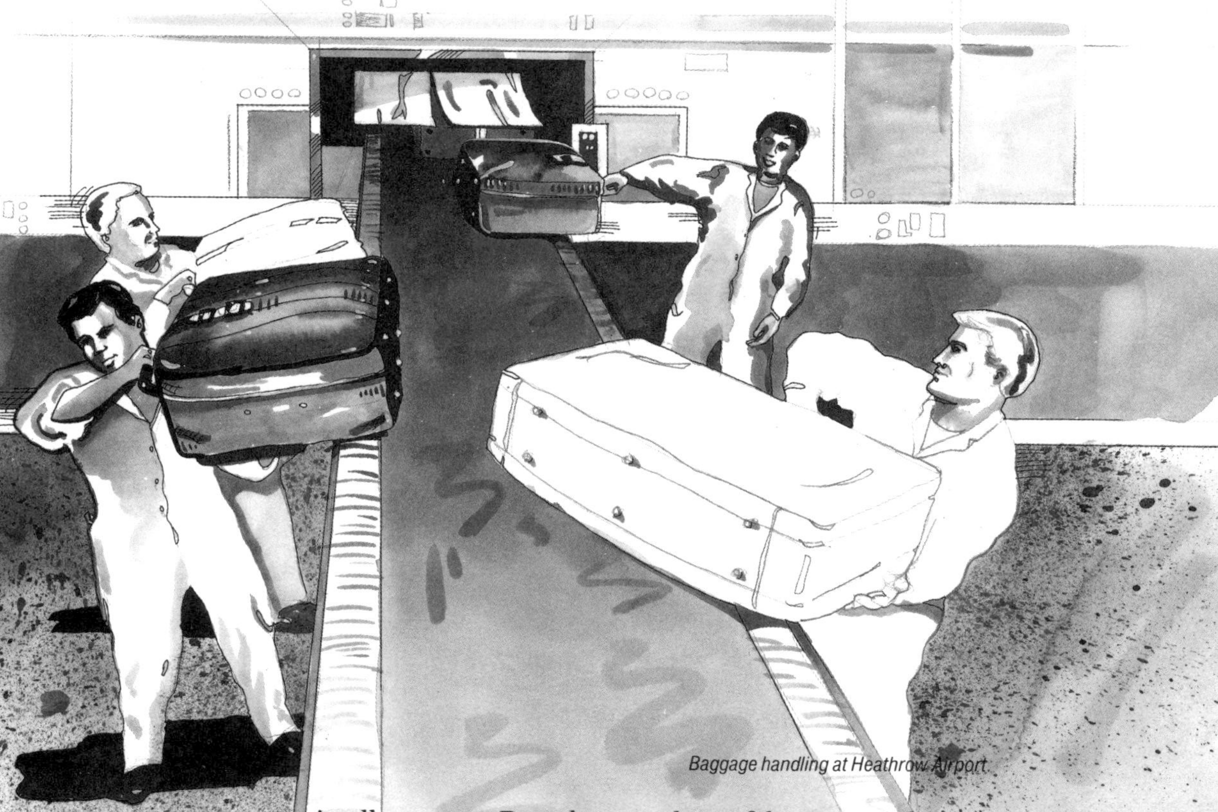
Baggage handling at Heathrow Airport.

over a passenger's allowance. But the number of bags can change up to the last 75 minutes before a flight. The airline is selling space and the courier companies only make estimates of their totals. The actual numbers are not known until the overloaded 'passengers' appear at check-in and all the bags are counted. Fifteen minutes to go and one courier still has not arrived.

The couriers can be old or young, somebody's mother or somebody's friend. They are a legality. On arrival at each destination they must move fast to check the bags through customs and hand them on to their company's representative. Then back on the aeroplane to accompany their remaining bags to the next stop. Qantas allocates couriers the same seats on every flight and cabin crew move them quickly to the front on arrival at a transit stop.

EBQ will carry some mail tonight, but it will be nowhere near the weight of the courier bags. The national carrier has the mail contract. British Airways passes some mail on to Qantas, particularly non-urgent, bulky parcels. Eight hundred kilograms of packages and parcels go on board, topping up containers or stored loose in bulk hold 5.

Down in the baggage handling area lone suitcases appear, upright on the moving belt, looking curiously vulnerable. The Duty Officer checks the baggage being stowed in containers — the exact numbers being noted on the outside, for load control — and

lifts up the odd grubby courier bag, to spot check its weight. Sometimes destination tags get ripped off passengers' suitcases on the journey down the conveyor belt. An old tag left on from a previous flight can be read in error and the suitcase despatched across the wrong ocean. Sometimes, as the baggage handler lifts a bag to stow into the container behind him, the container is moved and another put in its place. The bag is stowed, but this container has a new destination.

Passengers who have booked in at check-in desks go upstairs to endless activity if they wish to and have the time. They can telephone anywhere in the world. They can eat, drink, buy lambswool sweaters, tartan skirts, Royal Worcester china, swimsuits, watches, books, decorated tins of tea. They can post letters or stock up on medicines. The artificial mini-city of Heathrow is available to passengers and anyone else at the airport. But once past desks where their flight documents are checked to see that they are bona fide passengers, and through passport control and security, the spaces are less crowded. Only passengers can sit here in the departure lounge amongst the rows of orange seats, eat and drink at these restaurants, and spend money at this second line of shops, which includes large areas of duty-free and tax-free temptations.

The number of passengers who have reserved seats on QF2 for each class is known. Some will never appear. They are 'noshows' — people who have booked on two flights at once, who fail to cancel, who have missed a connection or cut it too fine getting out to the airport. Some passengers cancel seats so late that their seats can't be resold. Qantas, like all airlines, overbooks. Overbookings are meant to offset noshows and late cancellations. Calculating the amount of overbooking prudent for each route at each time of the year is a specialised job, known as capacity planning. Around one in 20 of all Qantas reservations ends up as a noshow.

Most flights carry airline personnel allowed to fly either on a concession or free. Thirty are hoping to board tonight's QF2 and have booked in at the 'stand-by' desk. They are the off-loadables, the cushion in case overbookings and noshows don't match. Airlines would lose a great deal of money if they did not overbook, but statistically only one in every 9000 Qantas passengers with a reserved seat fails to find one available.

The Flight Operations Officer has negotiated a slot time of 9.30 pm local time for QF2 with the air traffic controllers. QF2 must now be airborne within six minutes of the slot time, by 9.36 pm local time, 2036 UTC, or it will lose the slot and have to wait at the end of the queue. The slot time relates to the other traffic on the route and how busy the skies are. In the summer months

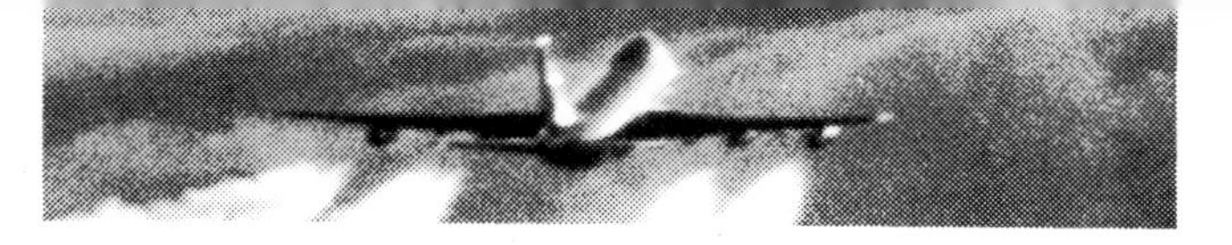

northern Europeans fly south to the sun on charters at night, causing much cross-traffic. Very occasionally, so little is happening that a slot time is not required.

Only one route is planned out of London for QF2, across the crowded European skies and down over the troubled Middle East and Gulf area. Doglegging along the airways, avoiding this, taking that into account, the sections of the route and their waypoints have been established. The Flight Operations Officer calculates the time it will take EBQ to fly between each waypoint, given the altitudes it is hoped the aeroplane will reach. He calculates the amount of fuel needed to fly each stage, given his knowledge of local conditions and the constantly updated meteorological information. Weather can change quickly in this part of Europe. To the 'trip fuel' he adds contingency fuel, which is enough for EBQ to reach a range of alternative airports like Sharjah or Muscat, and enough in case the Captain has to go into a holding pattern before being allowed to land at the alternative airport. For this London–Bahrain section tonight he adds an extra one and a half hours of fuel — around 13 tonnes.

The Flight Operations Officer organises the material for the tech. crew. Documents state the take-off runway (as indicated on the Heathrow computer), total distance EBQ will fly, other aircraft departing Heathrow in the same direction around the same time, and flight levels. There is a copy of the computer-stored flight plan as filed with air traffic control, and the important Nav Com Logs which give route information in printed form: latitude and longitude for waypoints, distances between waypoints, radio frequencies for calling up the air traffic controllers along the route. The Nav Com Logs form the flight's work sheet, to be filed on the flight deck and filled in en route.

A long, high, glass-topped counter with route maps of airways faces the door. The papers wait, neatly laid out. At five to eight local time the three pilots come in, weather-beaten faces, white shirts with gold stripes on the shoulders and wings on their chests. They look curiously out of their element, like beached seals: they are on their way to take control of the aircraft but they are not there; soon they will be in the air but they are still tied to ground requirements; they are not yet in command. The piles of paper are leafed through, the Flight Operations Officer proffers information, indicates the data provided, gives verbal confirmation of the written facts. The Captain discusses the meteorological information with the First and Second Officers, checks the Notams, looks at the alternatives, examines the Fuel Flight Plan and decides not to add any more fuel. Then they are gone. Planners have handed over to the doers.

In the passenger departure areas the noise of the 'other side' penetrates — a dense insistent sound of aircraft. There is a particular sweetish engine smell. Aeroplanes can be seen through the windows, but they are anonymous. Which belongs to which flight? Who knows? The announcements come. Singapore Airlines to Australia. British Airways flight BA11 to Australia. A passenger panics because his ticket says 'British Airways' and he has booked to fly with Qantas. Which flight is he supposed to be on? Ticket agents sometimes run out of an airline's tickets and use those of another airline. Any IATA carrier can issue a ticket for any other IATA carrier. It is the details filled in on the ticket which count, not the printer's ink.

At zero minus 75 minutes passengers for Qantas Flight Two are called to boarding gate number 28. People reach for their personal possessions and move along wide corridors, towards the finger of the departure terminal along which aeroplanes are loading. They are no longer people at an airport, or passengers for anywhere in the world. These are people who will travel on QF2 tonight at 9.15. Humans and their cabin baggage, they will be five per cent of EBQ's final take-off weight.

At the entrance to the boarding lounge passengers hand in their flight coupons and show their boarding cards. At check-in three hours ago, or 30 minutes ago, they were counted in sequence. Now they are ticked off against this sequence and the numbers build up — a headcount against check-in numbers. The airline would like to begin corralling as early as possible. Passengers wonder why they have this final wait, this final start-stop, at all. Any cabin baggage too bulky or too heavy is removed, for safety reasons, to be stowed in hold 5, the last to be closed.

By zero minus 60 minutes EBQ's own personal power unit is on, the Auxiliary Power Unit (APU) with its loud vacuum-cleaner whine. The air-conditioning system starts up. The second airbridge travels out. The cabin crew and technical crew will be on board within a few minutes. Vehicles cluster heavily along the cargo handling side, away from the airbridge and the fuel dispenser.

At zero minus 45 minutes people in wheelchairs go on board QF2. Then families with young children. The journey which has been beginning for some all day is at last beginning to happen.

Zero minus 30 minutes is 'flight closeout'. A line must be drawn. Things can be added to the aeroplane after this point but the timing gets difficult. A late passenger will have to pass through immigration and security, cover a physically long distance to the aeroplane and not fuss if, after all, he fails to board. His baggage has to be processed and loaded into the hold. Changes may be made to the load sheet until zero minus 10 minutes at the absolute

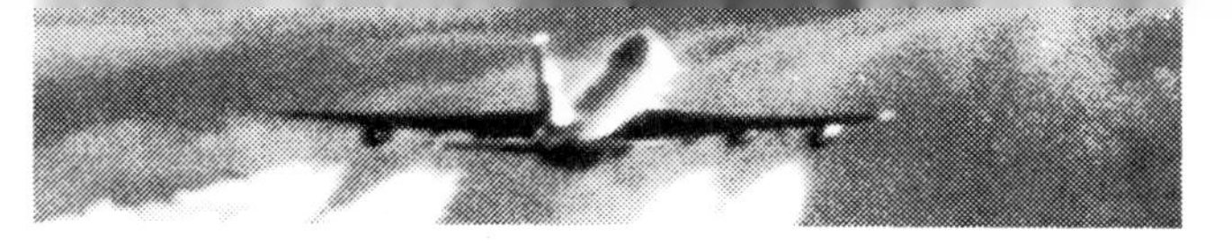

latest, but a completed load sheet has to be got up to the flight deck in time for the Captain to check and sign it. Zero minus 30 minutes is the crucial checkpoint for the Duty Officer. Cargo and passengers and food should all be in position, and fuelling nearing completion. The Duty Officer is like a sprinter on the blocks, ready in case something goes wrong.

Cargo door 1 is shut. Fuel is still coming on board. The passengers are settling and stand-by passengers are being allowed on. The service container still waits on its trolley, at an angle, like a piece of driftwood. Green, yellow and blue courier bags creep up the mobile conveyor belt into hold 5.

Zero minus 18 minutes. The service container goes up into cargo hold 4, position 44L, the priority baggage container arrives and is loaded into position 44R just by the door, and the hold is closed, the door swinging slowly downwards to fit the curve of the fuselage.

Zero minus 15 minutes. An engineer in white overalls walks away from the port wing carrying a milk crate of glass bottles. The water which accumulates in the fuel tanks from condensation has been drained off, and the bottles contain samples of fuel from each tank. The ground engineer will test the liquid by poking in a screwdriver smeared with a water-finding green paste. If it turns purple more impurities will need to be drained off. The ground engineer walks around the body of the aircraft, visually checking surfaces for any last minute knocks or damage to the fabric. He shines a torch through tiny glass portholes in EBQ's belly to check visually that the cargo doors are closed. Lights on the flight deck give the Flight Engineer the same information.

EBQ's wings are drooping with the weight of fuel, the jaunty upward line of empty wings converted to a slightly downturning curve. The wings have dipped a metre at the tips. The aircraft has sunk a little onto its haunches, the fuselage travelling 15 centimetres (six inches) down the landing gear legs.

The ground crew, who have been loading cargo and baggage, gather near the nose of EBQ, by the tug which will push it out. Big heavy men. A headset dangles, on a cord, from the nose wheel landing gear, so the supervisor can be in contact with the flight deck.

Slowly, subtly, the aeroplane is changing status. It has been done unto. Now it begins the doing. Vehicles have been parked hard up against it, pushing and manoeuvring things on board, hemming the body in. Two airbridges have fed passengers down its neck. Its tail end has hung open being stuffed with final cargo. Now its belly is full, its body is full, its wings are heavy with fuel, the flight deck is operational. The ground vehicles pull away,

Heathrow Airport control tower.

leaving the aircraft almost clear of ground-based matter. Number 2 airbridge retreats, lights flashing. The last things go into bulk hold 5 and a handler runs up the conveyor belt and closes the door. Ground power is disconnected and the pipes carrying the cables are folded back. Only the low, powerful tug keeps the old familiarity, but the coupling of tug to nose wheel is done gently, tow-bar connections made with respect. Two huge power sources joined at the beckoning of one finger.

A passenger is missing. Someone checked in but hasn't boarded. The lists are gone through at speed, the passenger's name and seat number identified, and she is found, sitting quietly on the aeroplane, unaware that she is 'lost'. As is usually the case, it is a clerical error.

The Captain is chafing to get away. The load sheet could not be completed until the exact number of passengers on board was

A fire tender hovers as OF2 awaits permission to push back from the parking bay

known. The Duty Officer makes sure that the load sheet is taken up to the flight deck for the Captain to check and sign, and reports an aircraft with loading complete.

EBQ begins to create its own noise. A hydraulic pump turns with hoarse grunts. A rush of air, and engine number 4 is ready to start. A small spurt of water dribbles from EBQ's centre belly down onto the concrete, like a bull relieving itself. The sun has gone but the evening is bright and clear. EBQ waits, no longer a passive container. It is active, about to have movement. A bright red light begins to revolve under the fuselage, another on top. The aircraft is alive. Warning is given to anyone on the ground. The airbridge pulls back. Cleaning ladies have moved into the boarding lounge.

Permission to start up has been granted from the control tower. The Captain checks by telephone with the ground engineer

that the doors are closed, that the pins locking the landing gear in the extended position are out, that all is clear to start engine number 4. The tarmac is not visible from the flight deck. The ground engineer reports that number 4 fan is turning. Engine number 4 comes on, loudly. A patrolling fire tender hovers on the perimeter. EBQ appears to throb, poised.

Commands and instructions pass through headsets, on the flight deck, in the control tower, down to the ground crew supervisor. Permission to push back is granted. Chocks are pulled out, brakes are off. A wave of the arm from the supervisor and the tug begins to push EBQ backwards, out of the parking bay. 21.15 hours. An on-time departure. A man walks at each wing tip, guarding the length and width of the aircraft. The flight deck gives no rear vision. Engine number 1 goes on. EBQ is being pushed backwards, steered into a turn so the nose faces the taxiway. It stops. Brakes on, chocks in place on the nose wheel. The tug is disconnected and pulls free. Engines number 2 and 3 come on and the ground engineer reports that the fans are turning. The First Officer is doing the take-off. He waits, fingers around the throttles. The bright cockpit lights go out and immediately the instrument panel lights take over, dominating the darkness. 'OK ground you can disconnect.' The headset is removed. The metamorphosis is almost complete. EBQ is no longer physically tied to anything on the ground. Clearance from the control tower to taxi.

'Qantas Two right turn on the green lights.' The engineer standing on the left checks that all is clear on the ground and waves visual clearance with two illuminated wands. Chocks off, brakes off, the First Officer throttles gently forward and EBQ taxis under its own power towards the runway. The pilot tests the feel of the control surfaces. Elevators flex down and back, the rudder swings, flaps dip and straighten.

The litany of a busy airport sounds through headphones. 'Alpha Fox push-back approved.' 'Two Zero Three push-back approved.' 'Qantas Two follow the greens.' The First Officer calls for the Before Take-off checklist, EBQ wobbles and rocks a little as it taxis on past lit hangars, other aircraft, across surfaces indented with lights. 'Qantas Two your airborne time is 36.' 'Dover One Foxtrot your squawk is 3372.' 'Qantas Two follow the Sabena.' EBQ lines up and stops. A final arching of shoulders under shoulder restraints, adjusting knees, hands waiting on controls.

'Qantas Two cleared for take-off.' V_1 rotate V_2, and EBQ is airborne and climbing in a long curve. It is 9.29 pm local time, 2029 UTC. This is the 396th departure Heathrow control has dealt

with today. The multitudinous lights below begin to cluster, islands of light in blackness, then assume neutral distance. Over the Channel at 29,000 feet, EBQ, under European control from Maastricht, Holland, goes on to automatic, and the First Officer shifts his body free from the controls. A snack is brought in for the tech. crew. Across the French coast at 9.52 pm, 2052 UTC, and EBQ turns towards the right on a new heading. The Second Officer leaves for his break. Paperwork is being got through. The tech. crew discuss air traffic control.

Bahrain will be reached early tomorrow morning local time, Singapore tomorrow evening local time, the coast of Australia will be crossed at 00.39 hours on Monday Western Australian time, Sydney will be reached early on Monday morning and Melbourne at 09.05 hours local time. EBQ will have made 2098 landings and flown 46 hours 35 minutes block to block, Melbourne to London and back, Friday to Monday. And at 17.15 hours local time on Monday, EBQ will depart Melbourne for Sydney, Honolulu and San Francisco, as QF3.

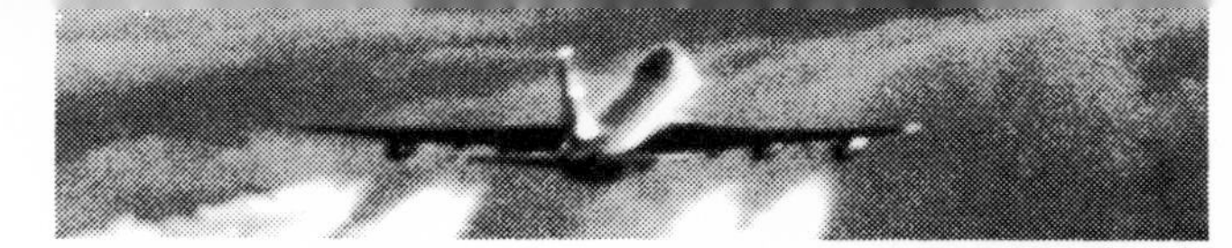

APPENDICES

APPENDIX 1

Airline Prefixes

The following list contains the prefixes used by a selection of airlines.

AC Air Canada
AF Air France
AI Air-India
AN Ansett Airlines of Australia
AZ Alitalia

BA British Airways
BN Braniff, Inc.
BR British Caledonian
BT Airlines of Northern Australia

CI China Airlines
CX Cathay Pacific Airways

DA Dan-Air Services
DH Tonga Air Service
DS Air Senegal
DZ Douglas Airways

EA Eastern Air Lines
EI Aer Lingus
EW East West Airlines

FJ Air Pacific
FO Western New South Wales Airlines

GA Garuda
GF Gulf Air
GH Ghana Airways
GJ Airlines of South Australia
GX Air Ontario
GZ Air Rarotonga

HA Hawaiian Airlines
HN NLM — Dutch Airlines
HX Holiday Express

IA Iraqi Airlines
IB Iberia
IC Indian Airlines

JL Japan Air Lines

KE Korean Air
KL KLM
KQ Kenya Airways
KU Kuwait Airways

LH Lufthansa
LO LOT — Polish Airlines
LY EL AL Israel Airlines

MH Malaysian Airline System
MS Egyptair
MV Airlines of Western Australia

NF Air Vanuatu
NZ Air New Zealand — Domestic

OA Olympic Airways
OS Austrian Airlines

PA Pan American World Airways
Pan Am Express
PC Fiji Air
PE People Express Airlines
PH Polynesian Airlines
PK Pakistan International Airlines
PR Philippine Airlines

QB Quebecair
QF Qantas Airways

RH Air Zimbabwe
RW Air Whitsunday

SA South African Airways
SK SAS — Scandinavian Airlines
SN Sabena
SQ Singapore Airlines
SR Swissair

TE Air New Zealand — International
TG Thai Airways International
TN Trans-Australia Airlines
TW TWA — Trans World Airlines

UA United Airlines
UB Burma Airways Corporation

VS Virgin Atlantic Airways
VT Air Polynesie

WN Southwest Airlines (USA)
WT Nigeria Airways
WX Air New South Wales

XE South Central Air
XV Mississippi Valley Airlines

YC Alaska Aeronautical Industries

ZH Royal Hawaiian Air Service
ZV Air Midwest

2B British Caribbean Airways

3P Pioneer Airlines

APPENDIX 2

The Phonetic Alphabet (International Aviation Code)

A Alpha
B Bravo
C Charlie
D Delta
E Echo
F Foxtrot
G Golf
H Hotel
I India
J Juliet
K Kilo
L Lima
M Mike
N November
O Oscar
P Papa
Q Quebec
R Romeo
S Sierra
T Tango
U Uniform
V Victor
W Whisky
X X Ray
Y Yankee
Z Zulu

APPENDIX 3

Aircraft Nationality and Registration Marks

Afghanistan **YA**
Albania **ZA**
Algeria **7T**
Andorra **C3**
Angola **D2**
Antigua and Barbuda **V2**
Argentina **LV**
Australia **VH**
Austria **OE**
Bahamas **C6**
Bahrain **A9C**
Bangladesh **S2**
Barbados **8P**
Belgium **OO**
Belize **V3**
Benin **TY**
Bermuda **VR-B**
Bhutan **A5**
Bolivia **CP**
Botswana **A2**
Brazil **PP/PT**
British Virgin Islands **VP-LVA/ZZ**
Brunei **V8**
Bulgaria **LZ**
Burkina Faso **XT**
Burma **XY/XZ**
Burundi **9U**
Cameroon **TJ**
Canada **C/CF**
Cape Verde Islands **D4**
Cayman Islands **VR-C**
Central African Republic **TL**
Chad **TT**

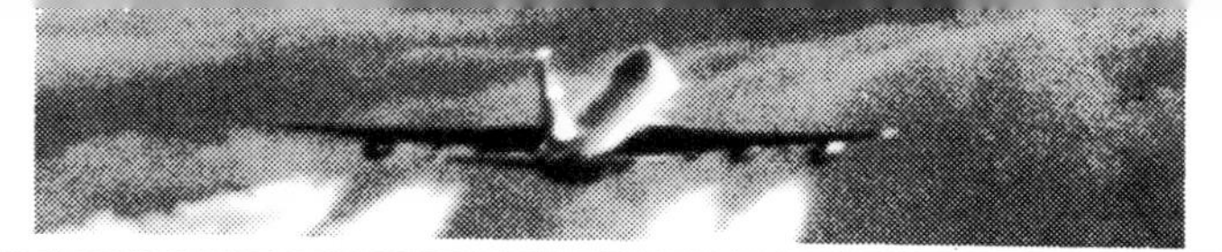

APPENDICES

Country	Prefix
Chile	CC
China (People's Republic)	B
China/Taiwan (R O C)	B
Colombia	HK
Comoros Islands	D6
Congo Brazzaville	TN
Costa Rica	TI
Cuba	CU
Cyprus	5B
Czechoslovakia	OK
Denmark	OY
Djibouti	J2
Dominica	J7
Dominican Republic	HI
Ecuador	HC
Egypt	SU
Eire	EI
El Salvador	YS
Equatorial Guinea	3C
Ethiopia	ET
Falkland Islands	VP-F
Federal Republic of Germany	D
Fiji	DQ
Finland	OH
France	F
French O. Depts/Protectorates	F-O
Gabon	TR
Gambia	C5
German Democratic Republic	DDR
Ghana	9G
Gibraltar	VR-G
Great Britain	G
Greece	SX
Grenada	J3
Guatemala	TG
Guinea	3X
Guinea Bissau	J5
Guyana	8R
Haiti	HH
Honduras	HR
Hong Kong	VR-H
Hungary	HA
Iceland	TF
India	VT
Indonesia	PK
Iraq	YI
Iran	EP
Israel	4X
Italy	I
Ivory Coast	TU
Jamaica	6Y
Japan	JA
Jordan	JY
Jordanian-Iraqi Coop Treaty	4YB
Kampuchea	XU
Kenya	5Y
Kiribati	T3
Korea (D.P.R.K.)	P
Kuwait	9K
Laos	RDPL
Lebanon	OD
Lesotho	7P
Liberia	EL
Libya	5A
Luxembourg	LX
Madagascar	5R
Malaysia	9M
Malawi	7QY
Maldives Republic	8Q
Mali	TZ
Malta	9H
Marshall Islands	MI
Mauritania	5T
Mauritius	3B
Mexico	XA/XB/XC
Monaco	3A
Mongolia	BNMAU
Montserrat	VP-LMA/LUZ
Morocco	CN
Mozambique	C9
Nauru	C2
Nepal	9N
Netherlands	PH
Netherlands Antilles	PJ
New Zealand	ZK/ZL/ZM
Nicaragua	YN
Niger	5U
Nigeria	5N
Norway	LN
Oman	A40
Pakistan	AP
Panama	HP
Papua New Guinea	P2
Paraguay	ZP
People's Dem. Rep. of Yemen	70
Peru	OB
Philippines	RP
Poland	SP
Portugal	CR/CS
Qatar	A7
Republic of Korea	HL
Romania	YR
Rwanda	9XR
San Marino	T7
Sao Tome	S9
Saudi Arabia	HZ
Senegal	6V/6W
Seychelles	S7
Sierra Leone	9L
Singapore	9V
Solomon Islands	H4
Somalia	60
South Africa	ZS/ZT/ZU
Soviet Union	CCCP
Spain	EC
Sri Lanka	4R
St Kitts-Nevis	VP-LKA/LZ
St Lucia	J6
St Vincent	J8
Sudan	ST
Suriname	PZ
Swaziland	3D
Sweden	SE
Switzerland and Liechtenstein	HB
Syria	YK
Tanzania	5H
Thailand	HS
The Vatican	HV
Togo	5V
Tonga	A3
Trinidad and Tobago	9Y
Tunisia	TS
Turkey	TC
Turks and Caicos Islands	VQ-T
Tuvalu	T2
Uganda	5X
United Arab Emirates	A6
United Nations Org.	4U
Uruguay	CX
USA	N
Vanuatu	YJ
Venezuela	YV
Vietnam	XV
Western Samoa	5W
Yemen Arab Republic	4W
Yugoslavia	YU
Zaire	9Q
Zambia	9J
Zimbabwe	Z

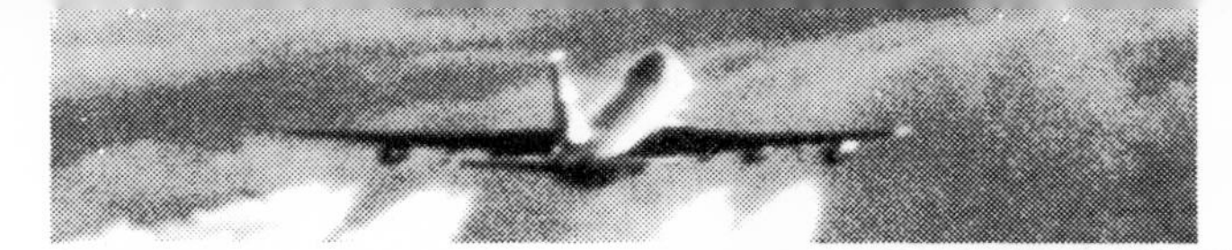

APPENDIX 4

City Codes

There are hundreds of international three-letter codes for aircraft destinations. The following list is a very short selection.

ADL Adelaide, South Australia
AKL Auckland International Airport, New Zealand
ALG Houari Boumediène Airport, Algiers, Algeria
AMS Schiphol Airport, Amsterdam, Netherlands
ATH Hellinikon Airport, Athens, Greece
BAH Bahrain
BEG Surcin Airport, Belgrade, Yugoslavia
BEY Beirut International Airport, Lebanon
BKK Don Muang Airport, Bangkok, Thailand
BNE Brisbane, Queensland
BRU Brussels National Airport, Belgium
BSB Brasilia, Brazil
BUD Ferihegy International Airport, Budapest, Hungary
BUE Ezeiza International Airport, Buenos Aires, Argentina
BUH Otopeni Airport, Bucharest, Romania
CAI Cairo International Airport, Egypt
CDG Charles de Gaulle Airport, Paris, France
CPH Kastrup Airport, Copenhagen, Denmark
DRW Darwin, Northern Territory
FRA Frankfurt-am-Main Airport, Federal Republic of Germany
GIG Rio de Janeiro International Airport, Brazil
HAN Hanoi, Vietnam
HBA Hobart, Tasmania
HKG Kai Tak Airport, Hong Kong
HLP Halim Perdanakusuma Airport, Jakarta, Indonesia
HNL Honolulu International Airport, Hawaii, USA
IAD Dulles International Airport, Washington, DC, USA
JFK John F. Kennedy International Airport, New York, USA
KUL Subang International Airport, Kuala Lumpur, Malaysia
KWI Kuwait International Airport
LAX Los Angeles International Airport, USA
LHR Heathrow Airport, London, UK
MAD Barajas Airport, Madrid, Spain
MEL Melbourne International Airport, Tullamarine, Victoria
MEX Mexico City International Airport, Mexico
MIA Miami International Airport, USA
MNL Manila International Airport, Philippines
NRT Narita Airport, Tokyo, Japan
ORD O'Hare International Airport, Chicago, USA
PEK Beijing (Peking), People's Republic of China
PER Perth, Western Australia
PHL Philadelphia International Airport, USA
PRG Ruzyně Airport, Prague, Czechoslovakia
ROM Leonardo da Vinci Airport, Rome, Italy
SDA Saddam International Airport, Baghdad, Iraq
SFO San Francisco International Airport, USA
SGN Ho Chi Minh City, Vietnam
SIN International Airport, Singapore
STO Arlanda International Airport, Stockholm, Sweden
SVO Sheremetyevo Airport, Moscow, USSR
SYD Kingsford Smith Airport, Sydney, Australia
TLV Ben Gurion International Airport, Tel Aviv, Israel
TPE Chiang Kai-shek International Airport, Taipei, Taiwan
VIE Schwechat Airport, Vienna, Austria
YMX Mirabel International Airport, Montreal, Canada
YYZ Toronto International Airport, Canada
ZRH Kloten Airport, Zurich, Switzerland

INDEX

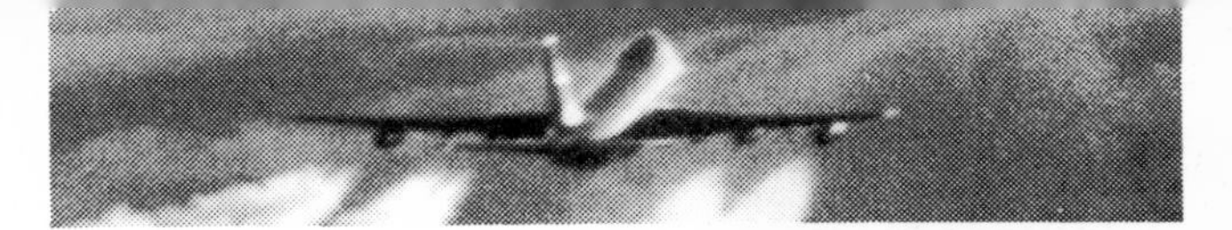

INDEX